T0233719

Boston Studies in the Philosophy and History of Science

Volume 315

Series editors
Alisa Bokulich, Department of Philosophy, Boston University, Boston, MA, USA
Robert S. Cohen, Boston University, Watertown, MA, USA
Jürgen Renn, Max Planck Institute for the History of Science, Berlin, Germany
Kostas Gavroglu, University of Athens, Athens, Greece

The series *Boston Studies in the Philosophy and History of Science* was conceived in the broadest framework of interdisciplinary and international concerns. Natural scientists, mathematicians, social scientists and philosophers have contributed to the series, as have historians and sociologists of science, linguists, psychologists, physicians, and literary critics.

The series has been able to include works by authors from many other countries around the world.

The editors believe that the history and philosophy of science should itself be scientific, self-consciously critical, humane as well as rational, sceptical and undogmatic while also receptive to discussion of first principles. One of the aims of Boston Studies, therefore, is to develop collaboration among scientists, historians and philosophers.

Boston Studies in the Philosophy and History of Science looks into and reflects on interactions between epistemological and historical dimensions in an effort to understand the scientific enterprise from every viewpoint.

More information about this series at http://www.springer.com/series/5710

Wenceslao J. Gonzalez

Editor

New Perspectives on Technology, Values, and Ethics

Theoretical and Practical

 Springer

Editor
Wenceslao J. Gonzalez
Faculty of Humanities
University of A Coruña
Ferrol, Spain

ISSN 0068-0346   ISSN 2214-7942 (electronic)
Boston Studies in the Philosophy and History of Science
ISBN 978-3-319-34389-1   ISBN 978-3-319-21870-0 (eBook)
DOI 10.1007/978-3-319-21870-0

Springer Cham Heidelberg New York Dordrecht London

Printed on acid-free paper

Springer International Publishing AG Switzerland is part of Springer Science+Business Media (www.springer.com)

Prologue
New Account for Technology and Its Relation to Values and Ethics

Wenceslao J. Gonzalez

Recent decades have witnessed an intense development in technology in many ways, mainly in the area of information and communication technologies (ICTs). These technological advancements have been followed by relevant philosophical analyses, which include new approaches to the role of values, in general, and ethical values, in particular. This new account of technology involves its being accepted as value-laden, instead of its characterization as value-free. In this regard, technology appears to be directly connected to ethical values, rather than being completely "neutral" from the ethical point of view.

Technology, Values, and Ethics focuses on a key issue today: the role of values in technology, with special emphasis on ethical values. This topic involves the analysis of internal values in technology (as they affect objectives, processes, and outcomes) and the study of external values in technology (social, cultural, economic, ecological, etc.). These values—internal and external—are crucial to the decision-making of engineers. In addition, they have increasing relevance for citizens concerned with the present and future state of technology, which gives society a leading position in technological issues.

Within this context, the book follows three main lines of research: (1) new perspectives on technology, values, and ethics; (2) rationality and responsibility in technology; and (3) technology and risks. This volume analyzes the two main sides involved here: the theoretical basis for the role of values in technology and a practical discussion on how to implement them in our society. Thus, the book is of interest for philosophers, engineers, academics of different fields, and policy makers. The style is appealing for a wide audience.

I   Contributions to New Perspectives on Technology, Values, and Ethics

Concerning the first part of the book, devoted to new perspectives on technology, values, and ethics, there are three papers: "On the Role of Values in the Configuration of Technology: From Axiology to Ethics," Wenceslao J. Gonzalez (University of A Coruña); "Values in Engineering and Technology," Ibo van de Poel (Delft University of Technology); and "Values Regarding Results of the Information and Communication Technologies: Internal Values," Paula Neira (University of Santiago of Compostela). They offer central elements of the general framework of the new account for technology and its relation to values and ethics.

In the context of the relevant improvements of technology in recent times, Wenceslao J. Gonzalez emphasizes that the analysis of the problem of values is a crucial topic. This involves considering what the role of values in the configuration of technology *is* and *ought to be*. The new scenario comes about for two main reasons: the frequent reflection on values regarding information and communication technologies (ICTs) and the acceptance nowadays of technology as value-laden, rather than value-free. Both aspects are involved in the reflection on ethical values. Thus, there is now an open door to new ethical analyses of technology as a human undertaking.

Following the idea of technology as value-laden, Gonzalez offers key elements for the new framework in three steps. (i) The values are related to the structural dimension of technology as a human enterprise and can clarify the dynamic perspective of this social undertaking. (ii) These values lead to an axiological analysis. This can consider the role of the "internal" values (those characteristic of technology itself) and the role of the "external" values (those around this human undertaking). (iii) The ethical values in technology, which are related to this free human activity, are also dual: endogenous and exogenous. Thus, they can be focused on this human undertaking related to the creative transformation of reality (endogenous perspective), or they can look at technology as intertwined to other human activities within a social setting (exogenous viewpoint). Accordingly, his philosophical analysis in Chap. 1 follows these three main steps.

Meanwhile, within this general setting of new perspectives, Ibo van de Poel discusses in Chap. 2 the role of values in engineering and technology. He argues that technology is not value-neutral as sometimes held. While values are regularly discussed with regard to technology, the role of values in engineering is far less researched. He distinguishes between two kinds of values that play a role in engineering. Internal values are values that are perceived by engineers as internal to engineering and engineering practice. This includes values like technological enthusiasm, effectiveness and efficiency, reliability, robustness, maintainability, and rationality. External values relate to effects of technologies developed by engineers outside engineering practice. Typical examples of external values are health and safety, human well-being, sustainability, and justice.

As Van de Poel argues, internal values are often conceived as ends in themselves (final values) by engineers, while they are instrumental values in a moral sense. The moral appropriateness of efficiency, for example, depends on the ends for which a technology is employed. An efficient killing device may be morally improper, while an efficient central heating system may be morally desirable. External values, like safety, health, and human well-being, are final values; however, to be effective in engineering, these values need to be internalized. This can be done, for example, through technical codes and standards, which translate these external values into often quite specific requirements for new technologies or through design approaches like design for X and value-sensitive design. While some external values like safety and health have already been internalized in engineering over time, this internalization is now occurring or still to start for some other external values like human well-being, sustainability, and justice.

Thereafter, in Chap. 3 Paula Neira goes a step forward and reflects on the need for assessment of technology aims, processes, and results. Her focus is on values regarding results of the information and communication technologies, which she analyzes from the perspective of internal values. Thus, her paper considers the internal values of information and communication technologies (ICTs), especially in the Internet and the World Wide Web. Among these values, the paper analyzes the degree of accessibility of the ICT, its versatility, its level of internal profitability or efficacy, and, obviously, its efficiency. Furthermore, economic values are analyzed from an internal aspect. These economic values have a role in the external level of this human undertaking but also have a less known role in the internal configuration of technology.

II   Dealing with Rationality and Responsibility in Technology

Regarding the second part of this volume, dealing with rationality and responsibility in technology, there are three papers: "Rationality in Technology and in Ethics," Carl Mitcham (Colorado School of Mines); "Knowledge and Moral Responsibility for *Online* Technology," Juan Bautista Bengoetxea (University of the Balearic Islands); "Risk, Uncertainty, and the Dimension of Technological Rationality," Amparo Gómez (University of La Laguna); and "Biotechnology, Ethics, and Society: The Case of Genetic Manipulation," Vicente Bellver (University of Valencia). They discuss different angles of the relation between rationality specific of technology and rationality characteristic of ethics.

Carl Mitcham offers a general approach to connections between rationality in technology and in ethics. Chapter 4 is structured in three main sections: (1) Rationality and Its Diverse Contents; (2) Engineering and Ethical Rationality; and (3) Ethics, Policy, and Rationality. It begins with the analysis of the language used regarding "rationality" as a way to clarify its meaning. Besides philosophical considerations on this issue, this paper takes into account the perspective of the social sciences. Regarding rationality, Mitcham discloses distinctions between instrumental and substantive rationality. In the case of ethics, he thinks of rationality and reasonableness. His view is that rationality is not a given in human affairs, but an end to be pursued. Thus, rationality is ultimately justified by appealing to some vision of the good.

Mitcham offers a historico-philosophical examination of how such a justification has been differentially manifested in the particular aspect of technology known as the "engineering profession." His focus is on engineers in the United States who have shifted from envisioning a professional obligation to pursue use and convenience in the built world in terms of corporate loyalty to the protection of public safety, health, and welfare. A second examination of efforts to supplement politics (power-based decision-making) with policy (scientific or rational-based decision-making) notes increasing efforts to formulate ethics for policy work and policies for the promotion of ethics. A concluding suggestion of his paper is that any rationality is ultimately dependent on its conceptualization in ethical terms.

A more specific analysis is developed by Juan Bautista Bengoetxea, who focuses on the practical contexts for information and knowledge. This sphere is emphasized by recent views in philosophy of science, which are oriented towards the study of knowledge as located in community practices. In this regard, of special interest is the context of *online* activity. In principle, knowledge is subject to ethics, norms, and values because of its social character. In the case of virtual knowledge, its ethical and social nature takes firm root in its practices. Bengoetxea tries to articulate the shift from the traditional notions of knowledge and justification—based upon concepts such as *evidence* and *perceptual access to data*—to a new inquiry on online knowledge and information practices according to reconstructed criteria. This task requires a reflection on the very notion of "online community" and an analysis on the possibility of implementation of ethical and normative codes in those communities.

Along these lines, in Chap. 5 Bengoetxea follows four steps. First, he examines some particular features of the Kantian and Hegelian philosophies regarding the social role of practices and their relationship to ethics. He seeks to conceptually establish the basis for a discussion on norms and values in online practices. Second, he considers the possibility of creating a philosophical realm focused on virtual epistemic practices that could allow him to *reformulate* the notions of knowledge and justification. Third, he focuses on people's behavior involved in online activities of information and knowledge. He thinks that it is necessary to reconstruct many traditional and customary values, norms, and ethical codes of non-online contexts. Fourth, he makes a plea for the usefulness of ethical codes in epistemic online practices.

Technological rationality is at the core of Amparo Gómez's analysis in the Chap. 6. She makes it explicit that technological rationality is not limited to internal considerations about means-ends in a neutral space (i.e., unaffected by economic, social, political, or moral factors), which can be seen in the consequences of the technological use as externalities. Moreover, when technology generates risk and uncertainty, there are important external considerations for technological rationality itself. Commonly, the technological rationality has been dominantly focused on efficiency and effectiveness, understood as the main values—even the only relevant values—in technological decisions and evaluations.

But Gómez stresses that technology has a social dimension and its effects have entered the public sphere (including the political sphere). In addition, technology is a social enterprise characterized by different perceptions, points of view, and values regarding technological phenomena. Thus, the interaction between its internal and external dimensions affects technological rationality. Consequently, it should take into account issues posed by the political and social dimensions of risk and uncertainty. Therefore, the concept of instrumental rationality for technology becomes too narrow. Indeed, technology needs a rationality of ends as a dimension of technological rationality. She considers that technological research and technological production cannot be totally separated from its uses and its consequences and that they cannot be separated from the ends pursued.

An interesting aspect connected with these ideas on rationality is the evaluation of technology, in general, and biotechnologies, in particular. Vicente Bellver makes clear that nowadays topics like assisted reproduction technologies, stem cell research, gene interventions, or cloning are part of the public talk. He thinks that there is a lot of hope in these biotechnologies, but, at the same time, he considers that there are concern and ethical worries about them. Thus, he has in mind issues such as "what should we do to use biotechnologies applied to human life for our personal and common good?" or "do we need any kind of ethical limits?" In order to answer these questions, he considers that citizens need a certain education: how these biotechnologies work, how they interact with society, and which ethical principles are suitable to decide what to do.

Thus, in Chap. 7, Bellver deals with these topics of this kind of technologies that are directly related to human life. In his paper, after presenting some of the most relevant and challenging advances in these biotechnologies, he focuses his attention on the way that biotechnology, government organizations, economy, and the law interact in society. Thereafter, he considers a very well-known challenge of biotechnology concerning human condition: human enhancement by means of germline genetic modification.

III   Theoretical and Practical Orientations on Technology and Risks

In the third part of this book, oriented towards technology and risks, there are three papers: "Risk and Trust in Institutions that Regulate Technology: Challenges for a Socially Legitimate Risk Analysis," Hannot Rodríguez (University of the Basque Country); "The Social Dimension of Technology: The Control of Chemical and Biological Weapons," Brian Balmer (University College London); and "Technology and Ecological Values: Confronting *Normal Waste* as Unavoidable Matter in Modern Society," Helena Mateus Jerónimo (School of Economics and Management of the

University of Lisbon [ISEG-UL] and Research Centre in Economic and Organizational Sociology [SOCIUS]).

Technology and risks are interrelated in Hannot Rodríguez's analysis. In Chap. 8 he points out that social distrust of institutions that regulate technological risks cannot be understood simply as being motivated by irrational impulse or prejudice. Dynamics of trust and distrust are in fact related to legitimate epistemological, ethical, and sociopolitical challenges to institutional risk analysis. These challenges, illustrated in the light of the controversial European regulation of agri-food biotechnology, are analyzed in three models of trust: (a) the competence model, (b) the cultural model, and (c) the relational model. The analysis shows that the public uptake of technological innovation depends on attending to legitimate social concerns about safety.

Nevertheless, Rodriguez thinks that the implication of risk governance measures (such as the adoption of precautionary and participatory policies in risk analysis) is arguably curtailed by the socioeconomic imperatives that guide strategic technological innovations like nanotechnology. Thus, he maintains that the legitimizing and transforming capacity of risk governance will remain limited, unless more profound decisions and practices are promoted with regard to safety of techno-industrial progress.

Brian Balmer looks directly at the social dimension of technology. In Chap. 9 he examines some conceptual problems facing the arms control community in their efforts to control chemical and biological weapons. He approaches the problem as a historian and sociologist of science, but in this chapter he invites philosophers of science to bring their unique disciplinary contributions to bear on the same issues. Philosophers have already contributed to discussions about the ethics of undertaking or censoring scientific research where anticipated or serendipitous knowledge could be applied for malign purposes. Balmer wants philosophy of science to make a greater contribution and suggests that terms such as tacit knowledge, ontology, and underdetermination and approaches such as feminist philosophy of science might help clarify some of the reasons why these weapons are so difficult to control.

Balmer provides a set of definitions and outlines the main efforts to rid the world of these weapons. He describes the key provisions of, and thinking behind, the international treaties that outlaw chemical and biological warfare. Rather than focus on the legal minutiae, he outlines various conceptual problems: defining these weapons, the problem of "dual-use" (where the same research can be used for either benign or malign purposes), the problem of distinguishing defensive from offensive research, the problem of verifying whether or not states have abided by the terms of the treaties, and attempts to make sense of the apparent deep-seated cultural taboo that is associated with chemical and biological weapons. He also considers the history of nerve gas and the history of biological warfare to illustrate some of these problems and how they give rise to conceptual issues that can be discussed using ideas and arguments from the philosophy of science.

Helena Mateus Jerónimo pays attention to a quite different aspect of technology: the relation to ecological values. Her focus is on the problem of waste as unavoidable matter in contemporary society. She considers that waste is a topic somewhat neglected by academics, especially when compared with the number of studies dedicated to consumerism. But waste is something that is omnipresent in all societies, both past and present. It is an inherent and inevitable condition of societies, which has come to stimulate social, cultural, economic, and technological responses.

Based on a sociohistorical analysis, Mateus Jerónimo shows that waste only started to be seen as a "problem" in a time of widespread production and consumption. This perception comes from a long cultural process of improving sensitivities, behaviors, mentalities, and new philosophical and medical convictions. These stimulated the development of hygiene and sanitary reforms in urban public spaces. Thereafter, with the emergence of environmentalism, waste has become associated with environmental decay. This includes a reflection on our options regarding resources and standards of development as well as some cultural attitudes towards consumption and the rights of future generations. For this author, when waste is seen as recyclable matter, it is not just a "problem": waste also is a resource that has an economic, aesthetic, and environmental value.

IV On the Present Interest in Technological Doings

Nowadays, from different perspectives, there is a particular interest in technological doings. The volume takes the philosophical approach as complementary to the empirical studies on technology, such as sociological research on technological undertakings. On the one hand, there are papers that pay attention to internal values in technology, and, on the other, several papers are clearly oriented to the external values in technology. From this angle, the analysis considers some aspects of the "social turn," and, as is usual in philosophy, it includes the component of critical attitude.

These features of the internal and external analysis of technology are considered in the chapters of this book. I would like to express my appreciation to the collaborators of this volume. They were speakers at the Conference of Technology, Values, and Ethics (*Jornadas sobre Tecnología, valores y Ética*),[1] organized at the University of A Coruña, Ferrol Campus, on 15–16 March 2012. In this regard, my acknowledgement to the organizations that gave their support to this academic activity: the Spanish Ministry of Science and Innovation, the City Hall of Ferrol, the University of A Coruña (especially to the Vice-Rector of the Campus of Ferrol and Social Responsibility), and the Society of Logic, Methodology, and Philosophy of Science in Spain.

My recognition to the persons that have cooperated in the original event: the local committee and the support team. In this regard, I would like to mention Dr.

[1] This was the title of the *XVII Conference on Contemporary Philosophy and Methodology of Science*, a series of workshops coordinated by the author of this prologue and editor of the present book.

Antonio Bereijo, who passed away on 26 February 2014. He had a very strong commitment with this conference. In addition, he backed the *Conferences on Contemporary Philosophy and Methodology of Science* for years.

My acknowledgement is also to Springer—in particular, to Lucy Fleet—for the interest in this topic. In addition, let me point out that I am grateful to Jéssica Rey and Amanda Guillan for their contributions to the edition of this book, which follows a line of academic work that began in 1996.[2] Finally, my gratitude to the Centre for Philosophy of Natural and Social Sciences of the London School of Economics, where a stay as research visitor has made possible a substantial part of the edition of this book.

London, 16 July 2014

---

[2] It was in that year that these conferences started at the University of A Coruña. They have been followed by a list of publications that is known as *Gallaecia: Studies in Contemporary Philosophy and Methodology of Science*.

Contents

Part I
New Perspectives on Technology, Values, and Ethics

Chapter 1
On the Role of Values in the Configuration of Technology: From Axiology to Ethics

Wenceslao J. Gonzalez

Within the context of the relevant improvements of technology in recent times, the analysis of the problem of values—what the role of values in the configuration of technology *is* and *ought to be*—appears as a crucial topic. While this relation between technology and values has certainly received attention in the past,[1] there is now an increasing interest in this connection, to the extent that it can be deemed to be a key issue. The new scenario has come about for several reasons, among which two stand out: the frequent reflection on values regarding information and communication technologies (ICTs) [cf. Ricci 2011], which includes attention to ethical reflections on this influential technological branch[2]; and the acceptance nowadays of technology as value-laden instead of being considered as value-free,[3] which opens the door to new ethical analyses of technology as a human undertaking.

Moreover, insofar as technology is value-laden, these values can lead us to the structural dimension of this human entreprise or can clarify the dynamic perspective of this social undertaking. These aspects should be open to an axiological analysis, which can consider the role of the "internal" values (those characteristic of technology itself) as well as the role of the "external" values (those around this human undertaking). In addition, the ethical values in technology—both endogenous and exogenous—will be at the focus of attention, due to their relevance for this free human activity.

---

[1] See in this regard the set of papers complied in Shrader-Frechette and Westra (1997). This is also the case a large number of the papers included in Hanks (2010).

[2] This attention to ethical issues concerning information and communication technologies appears clearly in Graham (1999). A more general perspective can be found in Van den Hoven and Weckert (2008), and in the Cambridge handbook: Floridi (2010).

[3] To some extent, there is a similitude between this variation and the explicit change in the case of science. Cf. Gonzalez (2013a).

W.J. Gonzalez (✉)
Faculty of Humanities, University of A Coruña, Dr. Vazquez Cabrera, w/n,
15.403-Ferrol, A Coruña, Spain
e-mail: wencglez@udc.es

© Springer International Publishing Switzerland 2015
W.J. Gonzalez (ed.), *New Perspectives on Technology, Values, and Ethics*, Boston Studies in the Philosophy and History of Science 315, DOI 10.1007/978-3-319-21870-0_1

Accordingly, the philosophical analysis here will follow three main steps. First, the characterization of the framework of values in technology based on the distinction between "structure" and "dynamics." This approach depends on the features of "technology" as being conceptually different from "science," which also requires taking into account what is "technoscience." Second, the role of values in the sphere of axiology of technology, where several kinds of values are involved (economic, social, cultural, ecological, etc.). Some of them can be seen as being "internal" to technology, whereas others might belong to the "external" sphere to this human entreprise. Third, the role of the specific values of ethics in technology, which includes the endogenous perspective and the exogenous viewpoint.

1.1   Technology and the Framework for Considering Values

An initial approach to technology is to take into account its etymology. In this regard, the term "technology" is a kind of knowledge, insofar as it is the *logos* (the doctrine or learning) of the *techné* (Cf. Mitcham 1980). But "techné" might be in the realm of "arts," when the main aim is to create beautiful objects, or in the sphere of "technics," when this human activity seeks to build up useful items. *Techné* is a practical activity based on knowledge of experiences of the past and the present, which follows certain rules to get artistic products or to produce tools for useful purposes.

Historically, technology appears as a social activity based on qualified knowledge, which is commonly developed in an intersubjective doing that achieves specific aims. But this human contribution to society goes beyond *techné* in many ways, among them three: (i) the kind of *knowledge* used (scientific as well as specific of technology), (ii) the complexity of the *human undertaking* developed (essentially creative in order to achieve an actual innovation), and (iii) the characteristics of the *product* obtained as a consequence of the undertaking (frequently, a new artifact that may have a officially registered patent).

Certainly the kind of knowledge used in *technology* is different from technics, due to the sophisticated aims to be achieved and the variety of values involved in the selection of the designs. Technology is a human undertaking that has higher aims than mere technics, because technology is oriented towards the creative transformation of the previous reality (natural, social, or artificial) using scientific knowledge as well as specific technological knowledge. This transformative process follows values when technology builds up a product that should be tangible. The product might be a noticeable change of nature (e.g., a tunnel), a new kind of social reality (e.g., a new social order in a country) or a visible artifact (e.g., an aircraft). Ordinarily, when the final product of technology is an artifact, it might be registered in a patent.[4]

---

[4]The issue of the registration and used of patents has important practical consequences. See in this regard Wen and Yang (2010).

This could hardly be the case in the final outcome of a technics or when a science is developed (included applied science).[5]

Following this analysis, it seems clear that *technology* is more than knowledge used in a transformative way to get a final product or artifact. I consider that "technology" includes a variety of components. (1) It has its own language, due to its attention to internal constituents of the process (design, effectiveness, efficiency, etc.) and external factors (social, ecological, aesthetical, cultural, political, etc.). (2) The structure of technological systems is articulated on the basis of its operativity, because technology should guide the creative activity of the human being that transforms nature, social reality, or artificial items. (3) The specific knowledge of the technological undertaking—know-how—is instrumental and innovative: this kind of knowledge seeks to intervene in an actual realm, to dominate it and to employ it in order to serve human agents and society. (4) The method used is based on an imperative-hypothetical argumentation.[6] Consequently, the aims are the key to making reasonable or rejecting the means used by technological processes.

(5) There are values regarding the aims chosen and accompanying the technological processes. These values could be internal (such as realizing the goal at the lowest possible cost) and external (social, political, ecological, etc.). They establish the conditions of viability of the possible technology and its alternatives. (6) The reality itself of the technological process is supported by social human actions, which are based on intentionality oriented towards the transformation of the surrounding reality.[7] (7) There are ethical values endogenous to technology, insofar as it is a free human activity, and there are also exogenous values to the aims, processes, and results of technology, because this is a human undertaking developed in a social milieu.

Hence, technology can be seen as a human activity oriented to obtain a *creative and transformative domain* of that reality—natural, social, or artificial—on which it is working. Primarily, technology does not seek to describe or to explain reality, because there is already a discovered reality, which is known to some extent (and its future can be predicted),[8] that technology wants to change according to certain

---

[5] Usually, the outcomes of science are public and of free access to users. However, the characteristic products of technology can be patented and, therefore, be initially private and with no free access for users.

[6] This imperative-hypothetical argumentation is different from the kind of argumentations commonly used in science, such as the hypothetical-deductive or the inductive-probabilistic.

In technology, if the aim is accepted, then the means and costs should be considered. If the means can guarantee the achievement of the aim—in a finite number of steps—and the estimated costs seem reasonable, then these means should be used to obtain the chosen objective, otherwise the "instrumental rationality," which is central in technology, does not work in this case. The "imperative" component—focused on the means that should be used—in technological argumentation comes after the acceptance of the hypothetical aims, means, and costs.

[7] These components are considered in Gonzalez (2005b), pp. 3–49; especially, p. 12.

[8] It is interesting the existence of institutions that explicitly supports a direct connection between technology and future. This is the case of Carnegie Mellon University in Pittsburgh, which has a "Hillman Center for Future-Generations Technologies."

aims. This technological domain appears at least in new designs and in the effectiveness-efficiency couple. Furthermore, it requires us to consider other aspects related to this human activity (ethical, economic, ecological, aesthetical, political, cultural, etc.).

Consequently, values play a role here. In one way or another, they point out what is worthy, or what has merit for us, either in objective terms, subjective terms (as individuals) or intersubjective terms (as a group or as a society). Thus, it might be the case that a technology could achieve its aims as such (i.e., effectiveness), but that we may consider it as non-acceptable from the point of view of other factors. Some values might be at stake here. They may be connected to economic criteria (e.g., the cost-benefit ratio), ethical principles (e.g., consent, fairness), aesthetical evaluations (e.g., beauty, harmony),[9] ecological effects (e.g., absence of pollution in the air or lack contamination of the rivers), political consequences (e.g., the civil liberties, the social progress) or repercussions for the dominant culture (e.g., in terms of compatibility regarding the shared criteria).

Meanwhile, the framework for considering values might be in the sphere of "technoscience."[10] But this term has been understood until now in rather different ways. (a) Technoscience is a new word that represents the *identity* between science and technology.[11] (b) Technoscience is an expression compatible with "science" and "technology," insofar as it expresses the sense of a strong practical *interaction* between science and technology while maintaining the difference in their references.[12] (c) Technoscience is the term for a new reality, a kind of *blend* or *hybrid* of science and technology.[13] (d) Technoscience could be just "techno-*logos*" or "techno-logic." This indicates that it is a subject that can be understood as directly based in science.[14]

Technoscience as *identity* between science and technology—the first option—includes that they have been strengthening their ties, and science and technology have got to the point where there are no semantic differences between both. In addition, they also have a common reference because there are no longer ontological differences between them. According to the practical interaction—the second position—the reference of technoscience is then twofold: there are two different aspects of reality that can have a *causal interaction* (or, at least, there is a relation which preserves the ontologies of science and technology).

---

[9] A notorious example of the search of the combination of art and technology is Steve Jobs. He was the cofounder of Apple, the founder of NeXT and the chairman of Pixar. He insisted in connecting aesthetical values and sophisticated technological procedures. Cf. Isaacson (2011), pp. 238–249, especially, pp. 239, 244 and 248.

[10] Cf. Latour (1987). See also Latour and Woolgar (1979/1986).

[11] This option is considered and criticized by Ilkka Niiniluoto, cf. Niiniluoto (1997a).

[12] Rescher defends a strong practical *interaction* between science and technology, even he does not commonly uses the term "technoscience." Cf. Rescher (1999), pp. 100–102. Rescher (1984/1999).

[13] Donna Haraway, "under her earlier figure of cyborg, sees technoscience as the full hybridization of science and technology," Ihde (2004), p. 121. See Haraway (1991).

[14] The analysis follows here Gonzalez (2005b), p. 9.

Along with the *hybrid* position—the third view on technoscience—the referent has properties that are different from science and technology.[15] In this case, the three of them can coexist (science and technology as well as technoscience). But techno-science understood as "techno-*logos*" or "techno-logic"—the fourth possibility— makes no sense for defining "technoscience:" the difference between "technics" and "technology" lies mainly in this point. *Technics* is practical knowledge of an accumulative kind, based on human experience but without the support of an explicit scientific knowledge; whereas *technology* is a human activity which transforms the reality (natural, social, or artificial) in a creative way, and it does so precisely on the basis of aims designed with the assistance of scientific knowledge (as well as by means of specific technological knowledge).

Conceptually, technology and science can be seen as different, even though they are heavily interwoven in many cases (primarily technology based on natural sciences, such as naval or aerospace technologies).[16] These conceptual differences can be noticed if science and technology are conceived around some constitutive elements, which include semantic, logical, epistemological, methodological, ontological, axiological, and ethical components.[17]

Sensu stricto, following those components, there is no genuine identity between science and technology. Thus, we can consider the theoretical reasons as well as the practical aspects to point out ontological differences between science and technology (cf. Niiniluoto 1997a, pp. 285–299; especially, pp. 287–291). This dissimilarity also involves a methodological distinction between scientific progress and technological innovation, even though this recognition is compatible with the acceptance of frequent cases of a strong practical *interaction* between science and technology (cf. Gonzalez 1997). These cases might be grounds to emphasize the use of the term "technoscience." But the existence of a causal interaction between science and technology should avoid two possible interpretations: the reduction of technology to a mere applied science, or the conception of science as a simple kind of by-product of technology.

1.1.1 Values in the Structural Dimension

Concerning the role of values in technology, the focus might initially be on the framework oriented towards the structural dimension or be led by the dynamic perspective. In the first case, the role of values is related to the configuration of

---

[15] *Technoscience* understood as "hybridization" or "symbiosis of science and technology" suggests examples, such as the interaction of the computer sciences and the information and communication technologies, which lead to products popularly called "new technologies," where the patents are on properties different from those obtained by previous technologies. See Echeverria (2003), pp. 64–68 and 71–72.

[16] These relations have been analyzed in many ways. They can be seen in a large number of publications mentioned in the bibliography of this chapter. Among them are Floridi (2004), part VII, pp. 305–349; and Olsen et al. (2009), part II, pp. 49–127.

[17] See, in this regard, Gonzalez (2005b), pp. 3–49; especially, pp. 8–13.

technology itself, i.e., values that belong to this social construction, which is different from other human activities (philosophy, science, arts, etc.). This analysis of the structural dimension involves taking into account a set of aspects, three of which are: (i) technology as a human knowledge, (ii) technology as a social undertaking oriented towards the creative transformation of reality; and (iii) technology as a product or artifact.

Unquestionably, technology is a human knowledge that needs to choose aims. This selection is made in order to develop processes that are oriented towards the achievement of concrete results. In this regard, the knowledge *that* ("descriptive"), the knowledge *how* ("operative") and the knowledge *whether* ("evaluative") are involved. In effect, technology requires some scientific knowledge, a specific technological knowledge (mainly concerning the artifacts), and the knowledge about what is preferable instead of that merely preferred. The latter is the sphere of values, which is related to evaluative rationality.[18] In this sphere of knowledge, values have a role related to the technological designs and the methodology used to developed such designs (e.g., economic values have a role in both steps) (cf. Gonzalez 1999a).

After technology as human knowledge, there is a more noticeable aspect for society: technology as a human undertaking developed in a social setting. This is a key feature in the comparison with science, because technology *creatively transforms* the reality. Thus, things are different when this human undertaking intervenes in nature, in society, or in the artificial world, because a new reality is eventually available (a tunnel, a bridge, an aircraft, a computer, a mobile phone, etc.). This aspect is also connected with instrumental rationality, which is a key factor in technology. In this practical realm, the role of values is mainly focused on means to achieve aims (which is commonly related to values such as effectiveness and efficiency). The values may also be economic (e.g., profitability in terms of cost-benefit). In addition, a set of values can be taken into account: ethical, ecological, aesthetical, sociological, cultural, political, etc.

Commonly, for the citizens, the most tangible aspect of technology is the product or artifact. Technology is, then, the reality available after the transformation of some elements of the world (natural, social, or artificial). Here again a set of values is involved. Some of them are related to the *item itself* that is available. In this regard, first, certain values might be purely instrumental or operative, such as utility; and, second, there is room for many additional values (aesthetic, cultural, sociological, etc.) in a product or artifact. These values are connected to the *setting* of such item, which is always historical. They might be of quite different kinds: social, economic, political, cultural, etc. Frequently, the technological product or artifact is registered in an official patent, which is used as a guarantee for its economic value for markets and organizations.[19]

Each of these three important approaches to technology—as knowledge, human undertaking, and product or artifact—involves two main categories of values according to its status: "internal" and "external." On the one hand, there are some

---

[18] Cf. Rescher (1999), pp. 79, 81, 92, and 172. See also Rescher (2003).

[19] On the difference between "markets" and "organizations," see Simon (2001).

values that are endogenous to the designs, processes, and results of the technology that is developed (effectiveness, efficiency, etc.); and, on the other, there exists some values that are exogenous to the contents of the technology as such, and these contextual values (ecological, social, cultural, political, etc.) complete the picture of the structural configuration of technology.

According to this analysis, when technology is seen in a structural dimension, there is a role for *internal* values and a role for *external* values. Their relation cannot be considered in terms of a rigid frontier or an axiological wall but rather as an interaction of values within a framework of holism of values. They can be considered as a sort of system where there is an interrelation between both sides, internal and external. Thus, although some values are mainly "internal" whereas other are mostly "external," there is a kind of osmosis between them. Both flanks of values require to taking into account that technology is not only structural but also dynamic.

1.1.2 Values in the Dynamic Perspective

Undoubtedly, there is a dynamic perspective regarding the role of values in technology. On the one hand, insofar as technology is a creative transformation of reality, it is always a dynamic enterprise. The set of aims, processes, and results (products or artifacts) sought by technology belong to a dynamic framework. On the other hand, innovation is a crucial factor for any technology. Commonly, an outdated technology is replaced by an innovative technology. Sometimes it precedes the actual demands of the users for the new products (see, in this regard, the innovative approach to information and communication technologies led by Steve Jobs) [cf. Isaacson 2011, *passim*; especially, pp. xx–xxi and 565–566].

Innovation is a characteristic feature of technology as such, and it might be in the aims, in the processes, or in the results of a specific technology (cf. Gonzalez 2013c). Hence, innovation can appear in the technological designs, in the human undertaking of making a technology, and in the final products of artifacts obtained. This innovation is always made according to some values, either internal or external. To be sure, the improvements of a technology can be based on endogenous technological values, such as effectiveness or efficiency, or can be built up on exogenous technological values (aesthetical, social, ecological, cultural, political, etc.).[20]

Both kinds of values—internal and external—have a role in the dynamic perspective on technology insofar as it is a *human undertaking*, and this trait requires the performance of agents seeking some aims. Thus, these are values related to technology as a *historical activity* that is due to agents with specific purposes. "Historical" is used here in a deep sense, connected to human beings and societies, which goes beyond the mere chronological dimension to embrace the possibility of radical changes, in addition to gradual changes or piecemeal modifications.

---

[20] According to Steve Jobs, "you can't win on innovation unless you have a way to communicate to customers," Isaacson (2011), p. 369.

Consequently, the technological variation can be richer than an "evolution," understood in terms of mere adaptation, in order to take in an actual facet of "historicity" in technology.[21] Therefore, it is open to the possibility of revolutionary changes, which can be recognized in some technological innovations (of which the Internet is one).

Historicity of technology is compatible with values seen in dynamic terms. Values can have at least a dual role in dynamic terms: on the one hand, they influence the technological changes in the three levels pointed out (in technology in general, in specific versions of it, and in the agents that build it up); and, on the other, the values themselves can be different over time, because of the emergence of new values, the modification of previous values, or the obsolescence of some values. Thus, besides the dynamic role on the technology (as knowledge, human undertaking, and product or artifact), there is another trait to be considered here: the change itself of the values related to technology.

How is this *change of values* possible? From the point of view of technology in general and specific forms of technology, it seems clear that there are *novelties*— above all, new realities—that are introduced by technology, mainly in some specific branches. These novelties can change the values accepted in a particular society. This has been the case with information and communication technologies (ICTs), because the Internet and the world wide web have oriented new values, internal as well as external. This change of values is based on new demands in public life, which is a combination of cultural elements (those constructed by human beings in a society) and natural traits (those grounded on what humans received). These aspects might give an account of the possibility of having historicity and objectivity of values regarding technology, in general, and in ICTs, in particular.

Regarding this issue of the variation of the values concerning technology, it seems clear that there are two main possibilities in *novelty*: a change "from within" and a modification "from outside." On the one hand, there might be an "internal" variation in technology, i.e., a change regarding the technological values already known. This possibility of variation can be considered in terms of new priorities or prestige of some values (such as ecological values for oil platforms, aesthetic values on phones and computers, or social values regarding roles in order to develop domotics) or the diminishing influence of the previous values, i.e., a minor consideration of something previously evaluated as worthy (such as the value of efficiency by all means, including what affects the protection of the natural environment). On the other hand, some values may arise due to the factor of novelty connected to the "external" context of historicity: human society is, by definition, historical and innovation leads to new technological realities, such as smart phones or tablets.

But the *change of values* should be seen from the angle of the agents. Regarding values in general, there are at least two large possibilities available from such a perspective: (1) values based on *human needs*, which commonly involve stability (and, in some cases, there might even be invariants); and (2) values based on *optional factors* of human life, which in principle include variation, insofar as they depend

---

[21] On the distinction on "process," "evolution," and "historicity," see Gonzalez (2013b).

on the degree of acceptance. The variation might be connected to time factors (such as a generational change) or other aspects (cultural, social, etc.).

First, values based on *human needs*, which make them strong, because the human needs give the values a solid ground, a place where is possible put down roots. There is then a support for the objectivity of those values,[22] a bedrock that is different from the subjective preferences or the intersubjective options of a community (either a small group or society as a whole). These values are those that basically remain the same over time, with some possible improvements due to an increase in the level of sophistication (e.g., related to clothing, housing, bridges, etc.).

Second, there are values based on *optional factors* of human life, which are related to the diversity of aspects of the life of persons and societies. This second kind of values may involve historicity in its content: (a) these human values are not Platonic entities to be shared by ontological participation; and (b) values are considered worthy according to some criteria (preferred, preferable, etc.), and their acceptance involves that they hold some merit. In this regard, the things considered as worthy by human agents can change from time to time (e.g., from a generation to the next) and even from one individual to another.

Once both possibilities of the agents are considered—values based on human needs and on optional factors—it seems to me that they can be used for the case of values regarding technology. Thus, there is a reason to think of the stability with some improvements in certain technologies (i.e., the refinements of current ones) and the clear variation of other specific technologies (i.e., the innovative new products).[23] The recognition of the existence of a historicity in the values does not involve *eo ipso* a relativism of a historical kind:

(I) The change in the values themselves or the variation in the level of acceptance of some values is commonly gradual or piecemeal instead of being instantaneous or fast. Thus, the change takes some time (e.g., ecological values). (II) There are few revolutionary changes of values, if we see them as holistic and in a very short period of time. They might happen after natural disasters or huge technological failures and, then, there is some objective basis for the changes of values (e.g., security). (III) The exogenous values usually require some intersubjective acceptance. Consequently, the content of values is frequently shared by a number of agents rather than by a single individual or small community.

1.2 Axiology of Technology

Subsequent to the acceptance of the presence of values in technology, both in the structural dimension and the dynamic perspective, the issue is then the roots of these values. The analysis can be made through axiology of technology, which is the philosophical study of values about technology (internal as well as external). This

---

[22] A relevant analysis of the objectivity of values is in Rescher (1999), ch. 3, pp. 73–96.

[23] On this issue, see section 3 "The Role of Innovation in Technology" in Gonzalez 2013c, pp. 19–24.

analysis can consider the "descriptive" aspect of values (which involves the recognition of the values actually used in technology) and the "prescriptive" facet on them (which implies that there are values that should have a leading role).[24]

Any axiological study should consider both sides, descriptive and prescriptive. Thus, axiology of technology should be dual in its initial focus, analyzing which values *are* actually in place in contemporary technology, and what *ought to* be done according to the accepted values. Commonly, when the analysis is made on values in general, instead of being focused specifically on ethical values, the philosophical study pays more attention to the "descriptive" aspect than to the "prescriptive" facet.

Prima facie, there are three possible levels of philosophical analysis about how this technological realm is built up: general, specific, and related to the agents. (i) Axiology of technology in general considers those values that can be in any form of technological expression. (ii) Axiology of a specific technology studies those values that belong directly to a concrete expression of technology: naval, aerospatial, mines, informative and communicative, electronic, etc. (iii) Axiology of the agents developing technology takes into account the values that are accepted by those that prepare the designs, choose the processes, and evaluate the results.

Even though the levels of axiology of technology in general and axiology of the agents developing technology are different, there is an interaction between them. Moreover, a source of innovation in technology comes from the interest of the agents in novelty. This might involve dissimilar or even diverse values between the standard technological values and the new values that come from creative agents, and they can lead to actual innovations. On the one hand, this can change the traditional conception of technology, commonly attached to "impersonal" or "abstract" values (effectiveness, efficiency, etc.), to a vision closer to human values of *our* undertaking (i.e., technology as *our* technology); and, on the other, the relevant technological innovations introduced by some agents, such as Steve Jobs, can bring about new values (e.g., new social and cultural values) in addition to economic values.[25]

However, there is a second kind of approach to axiology of technology, which is in place when the focus is on the *use of technology* instead of on how technology is built up. Thus, there is a distinction to be made between the *construction* of a technology and the *application* of a technology.[26] One thing is technology as a

---

[24] An interesting reflection can be made on the role of "technological imperatives," cf. Niiniluoto (1990).

[25] According to Walter Isaacson, Jobs "knew that the best way to create value in the twenty-first century was to connect creativity with technology, so he built a company where leaps of the imagination were combined with remarkable feats of engineering. He and his colleagues at Apple were able to think differently: They developed not merely modest product advances based on focus groups, but the whole new devices and services that consumers did not yet know they needed," Isaacson (2011), p. xxi.

[26] This distinction between construction and application can be seen in science: applied science is not the same as application of science. Cf. Niiniluoto (1993) and Niiniluoto (1995).

When applied science is developed, the aim is the solution to specific problems in a concrete realm of reality, whereas application of science is the use of that knowledge in a variable setting. Thus, using the same applied science, the applications of the available knowledge can be clearly different (for example, in hospitals).

human enterprise characterized by the constitutive elements pointed out already (language, system, knowledge, method, social undertaking, etc.), which commonly emphasizes three main aspects (the knowledge connected with the designs, the processes used to carried out them, and the products or artifacts obtained). Another thing is the use of technology in a variable setting, which is the practice of engineers or architects. This application is commonly developed in private organizations and public institutions. In this regard, it seems clear that, based on the same technology (i.e., the contents given by academic institutions), the practice of technology can vary from one person to another and from one place to another (even within the same city).

Through the practice of using technology some new values can be added. This is the ordinary case, because engineering or architecture are human activities developed in a social setting, within a historical context and economic support. Thus, the accepted values of the profession in engineering or architecture might be different according to cultural or historical factors. These contextual values can be quite diverse depending on the standards accepted in each society and the traditions of that professional community. Moreover, these aspects related to external values include some problems connected with the private organizations and public institutions that give economic support to the projects of engineers or architects.

1.2.1   The Role of "Internal" Values in Technology

Internal values are those that belong directly to technology itself or a specific technology (e.g., information technology), such as values regarding the design, the processes, and the results. They contribute directly to what technology *is* and *ought to be*. The values are "internal" insofar as they are endogenous for any technology or a particular version of it. Thus, they might be crucial for the possibility, operativity, and availability of a technology (communicative, naval, spatial, industrial, civil, mines, etc.). In addition, these values are commonly considered by the agents that build up technology. Thus, they can appear in the three axiological levels pointed out (general, specific, and agents).

At the same time, there are "external" values to technology. These are around the central technological factors already mentioned (aims, processes, and results), which are immediately connected with the constitutive elements of a technology (language, system, knowledge, methods, undertaking, etc). Thus, these values are exogenous insofar as they are related to the context of technology, such as the legal, social, cultural, political, ecological, or aesthetical aspects. But these external values to technology are also relevant, because they deal with what is worthy in many ways and what receives the attention of citizens, groups, societies, etc.

In principle, external values accompany internal values both for the structural dimension of technology and the dynamic perspective of technology. They are relevant for the configuration of technology as well as for its change over time. Due to its dynamic contribution, external values are open to possible changes. Thus, it is feasible that some of them might end up being "endogenous." This change occurs when an initially external value (such as the ecological value or the aesthetic value) become decisive for the design of technology (e.g., in technologies developed for protected areas of the world, such as Antartica, or in new smartphones, in order to have a more competitive design).

Likewise, there are values that can have an internal role as well as an external role. This is what happens with economic values (cf. Gonzalez 1999a). (i) They might be *internal* insofar as they intervene in the design and the methods. On the one hand, there are undeniable connections between the technological designs and the economist costs. Thus, technological knowledge requires considering economic values. And, on the other hand, these links affect the technological procedures, which need to consider the economic factors. (ii) Economic values can have an *external* role. Their presence is indisputable in the sphere of technology as human undertaking (e.g., wages to be paid, instruments to be used, business firms needed, etc.). But economic values also have a role in the sphere of technological policy, because they are considered in the decision-making of private organizations and public institutions (governments, international committees, etc.).[27]

1.2.2   The Task of "External" Values in Technology

Initially, there is a large number of "external" values related to technology: aesthetic, social, cultural, political, legal, economic, ecological, etc. Certainly, the aims, processes and results of technology have *tangible consequences* for the citizens, markets and organizations. The reason is clear: technology is oriented towards the creative transformation of the reality. Thus, its design looks to *change existing reality* (natural, social, or artificial) to produce new results. When the product is an artifact (airplane, automobile, computer, cell phone, tablet, etc.), the lives of the members of society can be directly affected. These changes might favor social development or they may be against the common good of citizens.[28]

External values can have a role in the three main stages of the technological doing. (1) They can intervene in the *design*, because technology uses scientific knowledge (know that), specific technological knowledge (know how), and evaluative knowledge (know whether). Thus, technology can take into account exogenous values (social, economic, ecological, etc.) in the design. This "external" task is clear in many technological innovations (smartphones, tablets, large airplanes, etc.), because they should consider the users of the product and the potential economic

---

[27] Commonly, this leads to legal aspects (international, national, and regional). In this regard, the precautionary principle has been discussed in many ways, as can be seen in the final bibliography of this chapter. Cf. D'Souza and Taghian (2010); Stirling (2006); and World Commission on the Ethics of Scientific Knowledge and Technology (2005).

[28] From time to time there are versions of Luddism and reflections on the problem. Cf. Glendinning (2003); Winner (2003); and Kitcher (2001), ch. 13, pp. 167–180.

profitability of the new artifact. (2) The technological *processes* are developed in public or private enterprises, which are organized socially according to some values (economic, cultural, political, etc.) and with an institutional structure (owners, administrators, etc.) (3) The final *result* of technology is a human-made product (commonly, an artifact) to be used by society, and it has ordinarily an economic evaluation in markets and organizations (cf. Gonzalez 2005b, 27–32).

Thus, insofar as technology is ontologically social as a *human doing*, it can be evaluated according to values accepted in the society. Furthermore, its product is commonly an item for society (even in the case of technology regarding nature, such as in the case of a tunnel). Moreover, the criteria of society have a considerable influence in promoting some kind of technological innovations (with their patents) or an alternative technology (with a new design, processes and product). Frequently, from the perspective of external values, technology is viewed with concern, especially in the case of recent phenomena (e.g., in accidents related to nuclear energy, the use of biotechnology with human beings, the nanotechnological risks, or in the dangers of new technologies such as hydraulic fracturing or "fracking").

These external values are very influential in the reflection on the *limits* of technology, when philosophy asks for the bounds (*Grenzen*) or ceiling of technology. This analysis of the terminal limits of technology should take into account the internal values as well as the external values (social, cultural, political, ecological, aesthetic, economic, etc.). In this regard, philosophy of technology considers the external values in the context of a democratic society interested in the well-being of its citizens,[29] thinking that their members can contribute to decision making (e.g., by means of associations or through the members of the parliament). The study of the limits of technology include the *prediction of* what technology can achieve in the future, but also require the *prescription* of what should be done according to certain values.[30] This prescriptive dimension of the external values of technology is more noticeable with there are clear risks for society at stake, either for the present or for the future (cf. Rescher 1983; Shrader-Frechette 1985, 1991, 1993).

Frequently, behind the analysis of values in technology, there are some influential philosophical orientations regarding what technology is and ought to be. Two of them seem to be especially important in the processes in technology and its results: (I) *technological determinism*, which assumes that the development of technology is uniquely determined by internal laws; and (II) *technological voluntarism*, which maintains that the change can be externally directed and regulated by the free choice of the members of the society. Both conceptions are related to the actual level of autonomy of human beings while doing technology.

On the one hand, technological determinists can argue that the development of technology—in general, and of a particular one—is *de facto* a complex system process where the imperatives have a role (at least, methodologically). On the other

---

[29] Cf. Niiniluoto (1997b). There is a relation between technological rationality and human happiness, cf. Rescher (1999), ch. 8, pp. 169–190.

[30] Prescription is attached to an evaluation and an assessment of the good and bad for society of the decision. This is a common practice in applied sciences such as economics, cf. Gonzalez (1998b).

hand, technological voluntarists can point out that the citizens do not have to obey *eo ipso* those imperatives. Ilkka Niiniluoto suggests an interesting middle ground between "determinism" and "voluntarism:" the commands of technology are always *conditional*, because they are based on some *value premises*. Thus, it is correct that we do not have to obey technological imperatives. Therefore, the principle that "can implies ought" is not valid insofar as not all technological possibilities should be actualized (cf. Niiniluoto 1990). Values should have a clear role here in decision-making regarding technology, which includes particular importance for the characterization of a "sustainable development" based on technology.[31]

1.3   Ethics of Technology

Among human values are the ethical ones. They belong to the core of ethics of technology, which deals with this human entreprise insofar as it is a free human activity. This involves a common distinction between "ethics" and "morals." Philosophically, *ethics* is related to the justification of human activity, according to some norms that can be based on principles that might have a universal form. Thus, there are important ethical systems throughout the history of philosophy (among them, the Aristotelian and Kantian proposals). Meanwhile, *morals* is conceived as the study of the actual way of behavior of individuals, groups and societies, trying to make explicit the rules and norms that they used *de facto* in one way or another.

When the focus of attention is on ethics, the philosophical study is more relevant insofar as it seeks a universal form or, at least, the widest level of generality. Obviously, there are many philosophical questions about technology and ethics that are relevant. Kristin Shrader-Frechette considers that they "generally fall into one of at least five categories. These are (1) conceptual or *metaethical* questions; (2) *general normative* questions; (3) *particular normative* questions about specific technologies; (4) questions about the ethical *consequences* of technological developments; and (5) questions about the ethical justifiability of various *methods* of technology assessment." (Shrader-Frechette 1997, p. 26).

Shrader-Frechette gives examples along these lines: (i) how ought one to characterize "free, informed consent" to risks imposed by a sophisticated technology?; (ii) are there duties to future generations potentially harmed by a technology?; (iii) should the US continue to export banned pesticides to other nations?; (iv) would development of a nuclear based energy (plutonium) technology threaten civil liberties?; and (v) does the benefit-cost economic analysis ignore noneconomic components of human welfare? It seems to me that here, again, there is the possibility of distinguishing three levels of analysis: general, specific, and related to the agents.

(a) Ethics of technology in general deals with those aspects that are relevant for any kind of technological enterprise. (b) Ethics of a specific technology takes care

---

[31] On this concept, see Niiniluoto (1994). On different aspects of this topic, see Mohapatra (2004); and Meyers (2012).

of the ethical problems in a concrete domain, such as the ethical issues of informative and communicative technologies (for example, in the case of the Internet or the social networks of the web). (c) Ethics related to the technological agents considers the ethical values used by them as criteria of *what is worthy*, as well as what they think *ought to be done* when they make designs, develop processes, and obtain results (product or artifacts). In this regard, the analysis goes beyond the mere morals of the technological agents (what they actually do nowadays) in order to offer an *ethical proposal* of what should these professionals do today and in the future.

Again there are two sides to the philosophical analysis of technology: the endogenous perspective and the exogenous viewpoint. The endogenous ethics of technology analyzes the steps of this free human entreprise, such as knowledge, human undertaking, and product or artifact. Meanwhile, the exogenous ethics of technology evaluates the contextual aspects of this human activity carried out in a social milieu. Thus, it takes into account ethical values socially assumed or institutionally accepted, which includes legislation at the different levels (regional, national and international) insofar as laws are embedded of ethical values.

1.3.1 The Endogenous Perspective on Ethics of Technology

As regards this endogenous perspective, it deals with the aims, the processes, and the results searched in technology. It is important do not think, in principle, of ethical values in mere terms of consequences but rather in terms of the ethical legitimacy.[32] Furthermore, the analysis cannot be made merely according to the legal standards in a country, because any ethical consideration of technology goes beyond such criteria. Thus, nobody can seriously consider as ethical some technologies that were commonly accepted in the past. Among them are machines and processes used in factories that were totally unhealthy due to the high level of pollution produced, even though those machines and processes of the factories were legal in many countries for years (and even now we can see examples of this phenomenon in several parts of the world).

Endogenous ethics of technology can start with *knowledge* insofar as it is not a mere "content" but rather an element of a human activity.[33] Initially, human knowledge as such—a cognitive or epistemic content—cannot be evaluated ethically (cf. Rescher 1999, pp. 159–162). However, knowledge in technology involves several aspects (scientific, specific of artifacts, and evaluative) that can be connected with the human undertaking of creative transformation of the reality. In this regard, knowledge in technology can be a part of a human free activity, and so it can be considered within an ethical setting. Moreover, knowledge is used for establishing the aims of the design, and these might be ethically acceptable or unacceptable.

---

[32] In addition, it is possible to think of the role of ethical ideals, cf. Rescher (1980); and Rescher (2009), part V, section 4, pp. 335–345.

[33] This is also the case in science, cf. Gonzalez (1999b).

Therefore, endogenous ethics can consider technology as a *human undertaking* that transforms reality (natural, social, or artificial). This social activity is, in principle, free in the original making (i.e., the innovative phase of creation of the technology) as well as in the practical used (i.e., the application to actual purposes by individuals or groups). Thus, the aims, processes, and results of this human activity of technology can be evaluated in ethical terms (i.e., the good and the bad, the right and the wrong, the correct and the incorrect, etc.).

Clearly, this is also the case in the use of technological expertise by engineers and architects,[34] whose moral performance can be ethically evaluated. In this regard, the single-minded solutions to technological problems should be avoided. Consequently, the ethical criteria can be considered for the possible technological alternatives, because there are commonly one or more technological alternatives to the technological undertaking in use. Those who use technology—mainly, engineers and architects—need to consider that ethics is endogenous to technological doing (mainly, because of the ethical evaluation of means and ends) and, therefore, ethics is not just exogenous (due to social or cultural pressure) to technology.

If the ethical evaluation of the human undertaking of technology is undeniable, due to the relevance of the characteristics of this free social activity (especially for the human person and the society as a whole), the presence of ethical values regarding the *product* or *artifact* might be clear as well. Sometimes the ethical evaluation can regard the artifact itself (e.g., a chemical weapon or a nuclear bomb), whereas in other cases the values of ethics can be focused on the use of the technological product or artifact. In this regard, there are several options, which includes the dual possibility: in some cases the utilization of a product might be good for the society, whereas in other cases the employment of that product might be harmful or noxious. This is what happens with some technological artifacts used in medicine.

1.3.2   The Exogenous Viewpoint of Ethics of Technology

Exogenous ethics has a role from the beginning insofar as technology is a human undertaking related to other human activities within a social setting. Technology is *our technology* in its structural dimension as well as in its dynamic perspective: it is a knowledge, an undertaking, and a product of human beings in society. (i) From the viewpoint of structure, the realm of technology are persons and groups willing to transform nature, society, or artifacts for social purposes, which might be ethically acceptable for that concrete milieu or unacceptable. (ii) Within the dynamic perspective, the exogenous values of ethical character in technology come under the influence of historicity: each historical period has variations in the evaluation of knowledge, undertaking, and product of human beings in society.

Given the relevance of the changes introduced by the technological transformations, the ethical criteria can be used in cases such as the "precautionary principle"—

---

[34] See in this regard Neely and Luegenbiehl (2008).

mainly, related to a reasonable sustainable development—[35] and other ethical contents assumed by laws (regional, national, and international). Thus, the ethical values might be important at different levels of society, and they have consequences for the developing countries. At the same time, the existence of a globalization involves that dynamic changes in technology are more intense than in the past. Private organizations (business firms, corporations, etc.) and public institutions are under regulatory conditions that should have ethical bases, such as preservation of environment, respect for people, avoidance of damage to the communities, etc.

Throughout the intense discussion on risks, where the exogenous ethical values of technology have a strong role, the existence of a variability from one country to another and from one historical moment to another seems clear. These ethical values—in the exogenous perspective—can be diverse depending on the ethical standards of each society, which includes the problems connected with public morality. Again, it affects the kind of technology accepted in a society, the specific forms of technology that might be considered at least as harmless, and the type of ethical principles accepted by the community of agents that build up a technology in a historical context.[36]

Here also appears the other side of the exogenous ethics: how society, when it is embedded in technology, can shape the agents that develop new technologies? Is the technology itself a source of ethical values? Regarding this issue, Carl Mitcham points out two options: "what might be termed *substantivism* is the position that technological change strongly shapes or influences social, political or human affairs; (…) as technology globalizes, socio-cultural orders converge. By contrast, *instrumentalism* views artifacts as tools that can reflect and be used in many different ways by diversity of human lifeworlds. (…) People shape their lives and cultures, then as individual or groups incorporate and adapt technologies in whatever ways they choose" (Mitcham and Waelbers 2009, p. 371).

The type of relation here may be two-way: on the one hand, technological innovation changes society (e.g., knowledge society under the influence of the Internet and the world wide web is clearly different from the society in the times of Great Depression), and these changes also shape ethical values (e.g., privacy, responsibility, solidarity, etc.); and, on the other hand, technology is completely human made (in its designs, in its undertakings, and its products), and its contents show a style of life chosen according to some social objectives. The assumption of the first direction makes the use of some criteria, such as utility, understood as an ethical principle for our society; but the relevance of the second direction emphasizes the role of human freedom regarding those tools that have been made in a society. Aims, processes, and results of technology are based on human decisions in a social setting.[37]

---

[35] On the relation between the precautionary principle and the sustainable development see McKinney and Hammer Hill (2001); and Som et al. (2009).

[36] See in this regard Shrader-Frechette (2005a); and (2005b).

[37] From the internal point of view, the methodology of technology has a central role. It is based on an imperative-hypothetical argumentation, where the aims are crucial to making reasonable or to

All in all, values have an important role for the structure of technology and for its dynamics over time. Internal values and external values are relevant for the designs, undertakings and products. Both sides — internal and external — are needed in order to clarify the values on which technology is built up and ought to be developed. Its configuration and historical dynamics depends on values.[38] Among these values, ethical ones have a very important role, both from an endogenous perspective and exogenous viewpoint. Ethical values can be considered in the knowledge, undertaking and products of technology. In addition, it seems clear that they have a role regarding aims, processes and results of this human activity. In this regard, society has the right to expect reasonable ethics of technology, and it should seek a rational technological policy for its citizens.

Bibliography

Achterhuis, H. (ed.). 2001. *American philosophy of technology: The empirical turn*. Translated by Robert Crease. Bloomington: Indiana University Press.

Adner, R., and R. Kapoor. 2010. Value creation in innovation ecosystems: How the structure of technological interdependence affects firm performance in new technology generations. *Strategic Management Journal* 31(3): 306–333.

Agassi, J. 1980. Between science and technology. *Philosophy of Science* 47: 82–99.

Agassi, J. 1982. How technology aids and impedes the growth of science. *Proceedings of the Philosophy of Science Association* 2: 585–597.

Agassi, J. 1985. *Technology*. Dordrecht: Reidel.

Agazzi, E. 1992. *Il bene, il male e la scienza. Le dimensioni etiche dell'impresa scientifico-tecnologica*. Milan: Rusconi.

Annavarjula, M., and R. Mohan. 2009. Impact of technological innovation capabilities on the market value of firms. *Journal of Information and Knowledge Management* 8(3): 241–250.

Aven, T. 2006. On the precautionary principle in the context of different perspectives on risk. *Risk Management* 8(3): 192–205.

Becher, G. (ed.). 1995. *Evaluation of technology policy programmes in Germany*. Dordrecht: Kluwer.

Berger, P.L., and T. Luckman. 1967. *The social construction of reality*. New York: Doubleday-Anchor.

Beyleveld, D., and R. Brownsword. 2012. Emerging technologies, extreme uncertainty, and the principle of rational precautionary reasoning. *Law, Innovation and Technology* 4(1): 35–65.

Bijker, W.E. 1994. *Of bicycles, bakelites, and bulbs: Toward a theory of sociotechnical change*. Cambridge, MA: The MIT Press.

Bijker, W.E., and J. Law. 1992. *Shaping technology/building society: Studies in sociotechnical change*. Cambridge, MA: The MIT Press.

Bijker, W.E., T.P. Huges, and T. Pinch. 1987. *The social construction of technological systems: New directions in the sociology and history of technology*. Cambridge, MA: The MIT Press.

---

rejecting the means used by the process of developing a technological artifact. And, from an external perspective, technology requires social values as human undertaking: the technological processes cannot be beyond social control.

[38] Historicity is important regarding some values in technology. The historical dynamics of technology requires to consider the evolutionary changes (the improvements in off-shore platforms, aircrafts, automobiles, …) and the "technological revolutions" (such as the computers). An analysis of the second ones is in Simon (1987/1997).

Borgmann, A. 1984. *Technology and the character of contemporary life: A philosophical inquiry.* Chicago: The University of Chicago Press.

Borgmann, A. 1999. *Holding on to reality: The nature of information at the turn of the millennium.* Chicago: The University of Chicago Press.

Bronson, K. 2012. Technology and values—Or technological values? *Science and Culture* 21(4): 601–606.

Bugliarello, G., and D.B. Doner (eds.). 1979. *The history and philosophy of technology.* Urbana: University of Illinois Press.

Bunge, M. 1966/1974. Technology as applied science. *Technology and Culture* 7:329–347. Reprinted in F. Rapp (ed.), *Contributions to a philosophy of technology*, 19–36. Dordrecht: D. Reidel.

Bunge, M. 1985. *Epistemology and methodology III: Philosophy of science and technology.* Dordrecht: Reidel.

Bynum, T.W., and S. Rogerson. 2004. *Computer ethics and professional responsibility.* Malden: Blackwell Publishing.

Byrne, E.F. (ed.). 1989. *Technological transformation.* Dordrecht: Kluwer.

Calluzzo, V.J., and Ch.J. Cante. 2004. Ethics in information technology and software use. *Journal of Business Ethics* 51(3): 301–312.

Camarinha-Matos, L.M., E. Shahamatnia, and G. Nunes (eds.). 2012. *Technology innovation for value creation proceedings.* New York: Springer.

Carolan, M. 2007. The precautionary principle and traditional risk assessment. *Organization and Environment* 20(1): 5–24.

Carpenter, S. 1983. Technoaxiology: Appropriate norms for technology assessment. In *Philosophy and technology*, ed. P. Durbin and F. Rapp, 115–136. Dordrecht: Reidel.

Christophorou, L.G., and K.G. Drakatos (eds.). 2007. *Science, technology and human values: International symposium proceedings.* Athens: Academy of Athens.

Clark, A. 2003. *Natural-born cyborgs: Minds, technologies and the future of human intelligence.* Oxford: Oxford University Press.

Clarke, S. 2005. Future technologies, dystopic futures and the precautionary principle. *Ethics and Information Technology* 7(3): 121–126.

Collins, H.M. 1982. *Frames of meaning: The sociological construction of extraordinary science.* London: Routledge and K. Paul.

Constant, E. 1984. Communities and hierarchies: Structure in the practice of science and technology. In *The nature of technological knowledge*, ed. R. Laudan, 27–46. Dordrecht: Reidel.

Cranor, C.F. 2004. Toward understanding aspects of the precautionary principle. *The Journal of Medicine and Philosophy* 29(3): 259–279.

Crombie, A.C. (ed.). 1963. *Scientific change: Historical studies in the intellectual, social and technical conditions for scientific discovery and technical invention. From antiquity to the present.* London: Heinemann.

Crousse, B., J. Alexander, and R. Landry (eds.). 1990. *Evaluation des politiques scientifiques et technologiques.* Quebec: Presses de l'Université Laval.

D'Souza, C., and M. Taghian. 2010. Integrating precautionary principle approach in sustainable decision-making process: A proposal for a contextual framework. *Journal of Macromarketing* 30(2): 192–199.

Das, M., and S. Kolack. 1990. *Technology, values and society: Social forces in technological change.* New York: Lang.

Dasgupta, S. 1994. *Creativity in invention and design.* Cambridge: Cambridge University Press.

Dasgupta, S. 1996. *Technology and creativity.* New York: Oxford University Press.

De Solla Price, D.J. 1965. Is technology historically independent of science? A study in statistical historiography. *Technology and Culture* 6:553–568.

De Vries, M.J., S.E. Hansson, and A. Meijers (eds.). 2013. *Norms in technology.* Dordrecht: Springer.

Dosi, G. 1982. Technological paradigms and technological trajectories: A suggested interpretation of the determinants and directions of technological change. *Research Policy* 11: 147–162.

Durbin, P. (ed.). 1987. *Technology and responsibility.* Dordrecht: Reidel.

Durbin, P. (ed.). 1989. *Philosophy of technology. Practical, historical and other dimensions.* Dordrecht: Kluwer.

Durbin, P. (ed.). 1990. *Broad and narrow interpretations of philosophy of technology.* Dordrecht: Kluwer.

Durbin, P., and F. Rapp (eds.). 1983. *Philosophy and technology.* Dordrecht: Reidel.

Echeverria, J. 2003. *La revolución tecnocientífica.* Madrid: FCE.

Elliott, B. 1986. *Technology, innovation and change.* Edinburgh: University of Edinburgh.

Ellul, J. 1954/1990. *La technique; ou, L'en jeu du siècle.* Paris: A. Colin, (2nd ed. revised, Paris: Economica, 1990.) Translated from the French by John Wilkinson with an introd. by Robert K. Merton: Ellul, J. 1964. *The technological society.* New York: Alfred A. Knopf.

Elster, J. 1983. *Explaining technical change: A case study in the philosophy of science.* Cambridge: Cambridge University Press.

Feenberg, A. (ed.). 1995. *Technology and the politics of knowledge.* Bloomington: Indiana University Press.

Feibleman, J.K. 1982. *Technology and reality.* The Hague: M. Nijhoff.

Fellows, R. (ed.). 1995. *Philosophy and technology.* Cambridge: Cambridge University Press.

Floridi, L. (ed.). 2004. *Philosophy of computing and information.* Oxford: Blackwell.

Floridi, L. (ed.). 2010. *Information and computing ethics.* Cambridge: Cambridge University Press.

Freeman, C., and L. Soete. 1997. *Economics of industrial innovation*, 3rd ed. Cambridge: The MIT Press.

Fuller, S. 1993. *Philosophy, rethoric, and the end of knowledge: The coming of science and technology studies.* Madison: University of Wisconsin Press.

García-Muina, F.E., E. Pelechano-Barahona, and J.E. Navas-López. 2011. The effect of knowledge complexity on the strategic value of technological capabilities. *International Journal of Technology Management* 54(4): 390–409.

Gehlen, A. 1965/2003. Anthropologische Ansicht der Technik. In *Technik im technischen Zeitlater*, eds. H. Freyer, J.Ch. Papalekas and G. Weippert. Düsserdorf: J. Schilling. Translated in abridged version as Gehlen, A. 2003. A philosophical-anthropological perspective on technology. In *Philosophy and technology: The technological condition*, eds. R.C. Scharff and V. Dusek, 213–220. Oxford: Blackwell.

Glendinning, Ch. 2003. Notes toward a Neo-Luddite Manifesto. In *Philosophy and technology: The technological condition*, ed. R.C. Scharff and V. Dusek, 603–605. Oxford: Blackwell.

Goldman, S.L. (ed.). 1989. *Science, technology, and social progress.* Bethlehem: Lehigh University Press (Coedited by Associated University Presses, London).

Gómez, A. 2001. Racionalidad, riesgo e incertidumbre en el desarrollo tecnológico. In *Filosofía de la Tecnología*, ed. J.A. López Cerezo, J.L. Luján, and E.M. García Palacios, 169–187. Madrid: Ed. OEI.

Gómez, A. 2002. Estimación de riesgo, incertidumbre y valores en Tecnología. In *Tecnología, civilización y barbarie*, ed. J.M. De Cozar, 63–85. Barcelona: Anthropos.

Gómez, A. 2003. El principio de precaución en la gestión internacional del riesgo. *Política y Sociedad* 40(3): 113–130.

Gómez, A. 2007. Racionalidad y responsabilidad en Tecnología. In *Los laberintos de la responsabilidad*, ed. R. Aramayo and M.J. Guerra, 271–290. Madrid: Plaza y Janés.

Gonzalez, W.J. 1990. Progreso científico, Autonomía de la Ciencia y Realismo. *Arbor* 135(532): 91–109.

Gonzalez, W.J. 1996. Towards a new framework for revolutions in science. *Studies in History and Philosophy of Science* 27(4): 607–625.

Gonzalez, W.J. 1997. Progreso científico e innovación tecnológica: La 'Tecnociencia' y el problema de las relaciones entre Filosofía de la Ciencia y Filosofía de la Tecnología. *Arbor* 157(620): 261–283.

Gonzalez, W.J. 1998a. Racionalidad científica y racionalidad tecnológica: La mediación de la racionalidad económica. *Ágora* 17(2): 95–115.

Gonzalez, W.J. 1998b. Prediction and prescription in economics: A philosophical and methodological approach. *Theoria* 13(32): 321–345.

Gonzalez, W.J. 1999a. Valores económicos en la configuración de la Tecnología. *Argumentos de Razón Técnica* 2: 69–96.

Gonzalez, W.J. 1999b. Ciencia y valores éticos: De la posibilidad de la Ética de la Ciencia al problema de la valoración ética de la Ciencia Básica. *Arbor* 162(638): 139–171.

Gonzalez, W.J. (ed.). 2005a. *Science, technology and society: A philosophical perspective*. A Coruña: Netbiblo.

Gonzalez, W.J. 2005b. The philosophical approach to science, technology and society. In *Science, technology and society: A philosophical perspective*, ed. W.J. Gonzalez, 3–49. A Coruña: Netbiblo.

Gonzalez, W.J. 2008. Economic values in the configuration of science. In *Epistemology and the social*, Poznan studies in the philosophy of the sciences and the humanities, ed. E. Agazzi, J. Echeverría, and A. Gómez, 85–112. Amsterdam: Rodopi.

Gonzalez, W.J. 2011a. Complexity in economics and prediction: The role of parsimonious factors. In *Explanation, prediction, and confirmation*, ed. D. Dieks, W.J. Gonzalez, S. Hartman, Th. Uebel, and M. Weber, 319–330. Dordrecht: Springer.

Gonzalez, W.J. 2011b. Conceptual changes and scientific diversity: The role of historicity. In *Conceptual revolutions: From cognitive science to medicine*, ed. W.J. Gonzalez, 39–62. A Coruña: Netbiblo.

Gonzalez, W.J. 2013a. Value ladenness and the value-free ideal in scientific research. In *Handbook of the philosophical foundations of business ethics*, ed. Ch. Lütge, 1503–1521. Dordrecht: Springer.

Gonzalez, W.J. 2013b. The sciences of design as sciences of complexity: The dynamic trait. In *New challenges to philosophy of science*, ed. H. Andersen, D. Dieks, W.J. Gonzalez, Th. Uebel, and G. Wheeler, 299–311. Dordrecht: Springer.

Gonzalez, W.J. 2013c. The roles of scientific creativity and technological innovation in the context of complexity of science. In *Creativity, innovation, and complexity in science*, ed. W.J. Gonzalez. 11–40. A Coruña: Netbiblo.

Graham, G. 1999. *The internet: A philosophical inquiry*. London: Routledge.

Habermas, J. 1968a/1971. *Erkenntnis und Interesse*. Frankfurt: Suhrkamp. Translated by Jeremy J. Shapiro. *Knowledge and human interests*. Boston: Beacon Press.

Habermas, J. 1968b. *Technik und Wissenschaft als "Ideology"*. Frankfurt: Suhrkamp.

Hacking, I. 1983. *Representing and intervening*. Cambridge, MA: Cambridge University Press.

Hacking, I. 1999. *The social construction of what?* Cambridge, MA: Harvard University Press.

Hanks, C. (ed.). 2010. *Technology and values: Essential readings*. Malden: Wiley-Blackwell.

Haraway, D. 1991. *Simians, cyborgs and women: The reinvention of nature*. New York: Routledge and Institute for Social Research and Education.

Heidegger, M. 1954/2003. Die Frage nach der Technik. In *Vorträge und Aufsätze*, ed. M. Heidegger, 13–44. Pfullingen: Günther Neske. Translated as Heidegger, M. The question concerning technology. In *Philosophy and technology: The technological condition*, eds. R. C. Scharff and V. Dusek, 252–264. Oxford: Blackwell.

Hiskes, A.L.D. 1986. *Science, technology, and policy decisions*. Boulder: Westview Press.

Hottois, G. 1990. *Le paradigme bioéthique: une éthique pour la technoscience*. Brussels: De Boeck-Wesmael.

Ihde, D. 1979. *Technics and praxis: A philosophy of technology*. Dordrecht: Reidel.

Ihde, D. 1983. *Existential technics*. Albany: State University of New York Press.

Ihde, D. 1991. *Instrumental realism: The interface between philosophy of science and philosophy of technology*. Indiana: Indiana University Press.

Ihde, D. 2004. Has the philosophy of technology arrived? A state-of-the-art review. *Philosophy of Science* 71(1): 117–131.

Ihde, D., and E. Selinger (eds.). 2003. *Chasing technoscience: Matrix for materiality*. Bloomington: Indiana University Press.

Isaacson, W. 2011. *Steve Jobs*. New York: Simon and Schuster.

Jacquette, D. (ed.). 2009. *Reason, method, and value. A reader on the philosophy of Nicholas Rescher*. Frankfurt: Ontos Verlag.

Jasanoff, S., G.E. Markle, J.C. Petersen, and T. Pinch (eds.). 1995. *Handbook of science and technology studies*. London: Sage.

Jaspers, K. 1958. *Die Atom-bombe und die Zukunft der Menschen*. Munich: Piper.

Jonas, H. 1979/1984. *Das Prinzip Verantwortung. Versuch einer Ethik für die technologische Zivilisation*. Frankfurt am Main: Insel. Translated as Jonas, H. *The imperative of responsibility: In search of an ethics for the technological age*. Chicago: The University of Chicago Press.

Kitcher, Ph. 2001. *Science, truth, and democracy*. Oxford: Oxford University Press.

Kopelman, L.M., D.B. Resnik, and D.L. Weed. 2004. What is the role of the precautionary principle in the philosophy of medicine and bioethics? *The Journal of Medicine and Philosophy. A Forum for Bioethics and Philosophy of Medicine* 29(3): 255–258.

Ladriere, J. 1977/1978. *Les enjeux de la rationalité: le defí de la science et de la technologie aux cultures*. Paris: Aubier/Unesco. Translated as Ladriere, J. *The challenge presented to culture by science and technology*. Paris: UNESCO.

Latour, B. 1987. *Science in action: How to follow scientists and engineers through society*. Cambridge, MA: Harvard University Press.

Latour, B., and S. Woolgar. 1979/1986. *Laboratory life: The social construction of scientific facts*. Princeton: Princeton University Press (2nd ed., 1986).

Laudan, R. (ed.). 1984. *The nature of technological knowledge: Are models of scientific change relevant?* Dordrecht: Reidel.

Lelas, S. 1993. Science as technology. *The British Journal for the Philosophy of Science* 44: 423–442.

Lenk, H. 2007. *Global technoscience and responsibility: Schemes applied to human values, technology, creativity and globalisation*. London: LIT.

Lowrance, W.W. 1986. *Modern science and human value*. New York: Oxford University Press.

Lowrance, W.W. 2010. The relation of science and technology to human values. In *Technology and values: Essential readings*, ed. C. Hanks, 38–48. Oxford: Wiley-Blackwell. (From Lowrance (1986), pp. 145–150.)

Lujan, J.L. 2004. Principio de precaución: Conocimiento científico y dinámica social. In *Principio de precaución, Biotecnología y Derecho*, ed. C.M. Romeo Casabona, 221–234. Granada: Comares/Fundación BBVA.

Lujan, J.L., and J. Echeverria (eds.). 2004. *Gobernar los riesgos. Ciencia y valores en la Sociedad del riesgo*. Madrid: Biblioteca Nueva/OEI.

Macpherson, C.B. 1983. Democratic theory: Ontology and technology. In *Philosophy and technology*, ed. C. Mitcham and R. Mackey, 161–170. New York: The Free Press.

Magnani, L., and N.J. Nersessian (eds.). 2002. *Model based reasoning: Science, technology, values*. New York: Kluwer Academic/ Plenum Publishers.

Maguire, S., and J. Ellis. 2010. The precautionary principle and risk communication. In *Handbook of risk and crisis communication*, ed. R.L. Heath and D. O'Hair, 119–137. New York: Routledge.

McKinney, W.J., and H. Hammer Hill. 2001. Of sustainability and precaution: The logical, epistemological and moral problems of the precautionary principle and their implications for sustainable development. *Ethics and the Environment* 5(1): 77–87.

Meyers, R.A. (ed.). 2012. *Encyclopedia of sustainability science and technology*. New York: Springer.

Michalos, A. 1983. Technology assessment, facts and values. In *Philosophy and technology*, ed. P. Durbin and F. Rapp, 59–81. Dordrecht: Reidel.

Mitcham, C. 1980. Philosophy of technology. In *A guide to the culture of science, technology and medicine*, ed. P. Durbin, 282–363. New York: The Free Press.

Mitcham, C. 1994. *Thinking through technology. The path between engineering and philosophy*. Chicago: The University of Chicago Press.

Mitcham, C., and R. Mackey. (eds.). 1983. *Philosophy and technology: Readings in the philosophical problems of technology*. New York: Free Press (1st ed., 1972).

Mitcham, C., and K. Waelbers. 2009. Technology and ethics: Overview. In *A companion to the philosophy of technology*, ed. J.-K. Berg Olsen, S.A. Pedersen, and V.F. Hendricks, 367–383. Malden: Wiley-Blackwell.

Mohapatra, K.M. (ed.). 2004. *Technology, environment and human values: A metaphysical approach to sustainable development*. New Delhi: Concept Publishing.

Myers, N. 2002. The precautionary principle puts values first. *Bulletin of Science, Technology and Society* 22(3): 210–219.

Neely, K.A., and H.C. Luegenbiehl. 2008. Beyond inevitability: Emphasizing the role of intention and ethical responsibility in engineering design. In *Philosophy and design: From engineering to architecture*, ed. P. Vermaas et al., 247–257. Dordrecht: Springer.

Niiniluoto, I. 1990. Should technological imperatives be obeyed? *International Studies in the Philosophy of Science* 4: 181–187.

Niiniluoto, I. 1993. The aim and structure of applied research. *Erkenntnis* 38: 1–21.

Niiniluoto, I. 1994. Nature, man, and technology – Remarks on sustainable development. In *The changing circumpolar north: Opportunities for academic development*, vol. 6, ed. L. Heininen, 73–87. Rovaniemi: Arctic Centre Publications.

Niiniluoto, I. 1995. Approximation in applied science. *Poznan Studies in the Philosophy of Science and Humanities* 42: 127–139.

Niiniluoto, I. 1997a. Ciencia frente a Tecnología: ¿Diferencia o identidad? *Arbor* 157(620): 285–299.

Niiniluoto, I. 1997b. Límites de la Tecnología. *Arbor* 157(620): 391–410.

Ojha, S. 2011. *Science, technology and human values*. New Delhi: MD Publications.

Olive, L. 1999. Racionalidad científica y valores éticos en las Ciencias y la Tecnología. *Arbor* 162(637): 195–220.

Olsen, J.K.B., S.A. Pedersen, and V.F. Hendricks (eds.). 2009. *A companion to the philosophy of technology*. Chichester: Wiley-Blackwell.

Ortega y Gasset, J. 1939/1997. *Ensimismamiento y alteración. Meditación de la Técnica*. Buenos Aires: Espasa-Calpe. Reprinted in Ortega y Gasset, J. *Meditación de la Técnica*. Madrid: Santillana.

Peterson, M. 2007. Should the precautionary principle guide our actions or our beliefs? *Journal of Medical Ethics* 33(1): 5–10.

Pinch, T.J., and W.E. Bijker. 1984. The social construction of facts and artefacts: Or how the sociology of science and the sociology of technology might benefit each other. *Social Studies of Science* 14:399–441. Published, in a shortened and updated version, as Pinch, T.J., and W.E. Bijker. 2003. The social construction of facts and artefacts. In *Philosophy and technology: The technological condition*, eds. R. C. Scharff and V. Dusek, 221–232. Oxford: Blackwell.

Pitt, J. (ed.). 1995. *New directions in the philosophy of technology*. Dordrecht: Kluwer.

Pitt, J. 2000. *Thinking about technology: Foundations of the philosophy of technology*. New York: Seven Bridges Press.

Radnitzky, G. 1978. The boundaries of science and technology. In *The search for absolute values in a changing world, Proceedings of the VIth international conference on the unity of sciences*, vol. II, 1007–1036. New York: International Cultural Foundation Press.

Rapp, F. (ed.). 1974. *Contributions to a philosophy of technology*. Dordrecht: Reidel.

Rapp, F. 1978. *Analitische Technikphilosophie*. Munich: K. Alber.

Regan, P.M. 2009. *Legislating privacy: Technology, social values and public policy*. Chapel Hill: The University of North Carolina Press.

Rescher, N. 1980. *Unpopular essays on technological progress*. Pittsburgh: University of Pittsburgh Press.

Rescher, N. 1983. *Risk: A philosophical introduction to the theory of risk evaluation and management*. Lanham: University Press of America.

Rescher, N. 1984/1999. *The limits of science*. Berkeley: University of California Press. Revised edition. *The limits of science*. Pittsburgh: University of Pittsburgh Press.

Rescher, N. 1999. *Razón y valores en la Era científico-tecnológica*. Barcelona: Paidós.

Rescher, N. 2003. *Sensible decisions. Issues of rational decision in personal choice and public policy*. Lanham: Rowman and Littlefield.

Rescher, N. 2009. The power of ideals. In *Reason, method, and value. A reader on the philosophy of Nicholas Rescher*, ed. D. Jacquette, 335–345. Frankfurt: Ontos Verlag.

Ricci, G.R. (ed.). 2011. *Values and technology*. New Brunswick: Transactions Publishers.

Rosenberg, N. 1994. *Exploring the black box: Technology, economics, and history*. Cambridge, MA: Cambridge University Press.

Sahal, D. 1987. *Patterns of technological innovation*. Reading: Addison-Wesley.

Sandin, P. 2004. The precautionary principle and the concept of precaution. *Environmental Values* 13(4): 461–475.

Sandin, P. 2009. A new virtue-based understanding of the precautionary principle. In *The ethics of protocells: Moral and social implications of creating life in laboratory*, ed. M. Bedau and E.C. Parke, 88–104. Cambridge, MA: The MIT Press.

Scharff, R.C., and V. Dusek (eds.). 2003. *Philosophy and technology: The technological condition*. Oxford: Blackwell.

Shrader-Frechette, K. 1983. Technology assessment and the problem of quantification. In *Philosophy and technology*, ed. P. Durbin and F. Rapp, 151–164. Dordrecht: Reidel.

Shrader-Frechette, K. 1985. *Risk analysis and scientific method: Methodological and ethical problems with evaluating societal hazards*. Dordrecht: Reidel.

Shrader-Frechette, K. 1991. *Risk and rationality: Philosophical foundations for populist reforms*. Berkeley: University of California Press.

Shrader-Frechette, K. 1992. Technology. In *Encyclopedia of ethics*, eds. L. C. Becker, and Ch.B. Becker, vol. 2, 1231–1234. New York: Garland Publishing. Reprinted also as Shrader-Frechette, K. 2010. Technology and ethics. In *Technology and values: Essential readings*, ed. C. Hanks, 60–64. Oxford: Wiley-Blackwell.

Shrader-Frechette, K. 1993. *Burying uncertainty: Risk and the case against geological disposal of nuclear waste*. Berkeley: University of California Press.

Shrader-Frechette, K. 1997. Technology and ethical issues. In *Technology and values*, eds. K. Shrader-Frechette and L. Westra, 25–31. Savage: Rowman and Littlefield (Originally published as Shrader-Frechette (1992)).

Shrader-Frechette, K. 2002. *Environmental justice: Creating equality, reclaiming democracy*. New York: Oxford University Press.

Shrader-Frechette, K. 2005a. Objectivity and professional duties regarding science and technology. In *Science, technology and society: A philosophical perspective*, ed. W. J. Gonzalez, 51–79. A Coruña: Netbiblo.

Shrader-Frechette, K. 2005b. How to reform science and technology. In *Science, technology and society: A philosophical perspective*, ed. W. J. Gonzalez, 107–132. A Coruña: Netbiblo.

Shrader-Frechette, K., and L. Westra (eds.). 1997. *Technology and values*. Savage: Rowman and Littlefield.

Simon, H.A. 1987/1997. The steam engine and the computer: What makes technology revolutionary. *EDUCOM Bulletin* 22(1): 2–5. Reprinted in Simon, H.A. 1997. *Models of bounded rationality*. Vol. 3: *Empirically grounded economic reason*, 163–172. Cambridge, MA: The MIT Press.

Simon, H. 1996. *The sciences of the artificial*, 3rd ed. Cambridge, MA: The MIT Press (1st ed., 1969; 2nd ed., 1981).

Simon, H.A. 2001. Complex systems: The interplay of organizations and markets in contemporary society. *Computational and Mathematical Organizational Theory* 7: 79–85.

Skolimowski, H. 1966. The structure of thinking in technology. *Technology and Culture* 7: 371–383. (Reprinted in Rapp, F. (ed.). 1974. *Contributions to a philosophy of technology*. Dordrecht: Reidel.)

Skolimowski, H. 1968. On the concept of truth in science and in technology. In *Akten des XIV. Internationalen Kongresses für Philosophie*, 553–559. Vienna: Herder.

Smith, C.A. 1980. Technology and value theory. *Proceedings of the Philosophy of Science Association* 2: 481–490.

Som, C., L.M. Hiltiy, and A.R. Köhler. 2009. The precautionary principle as a framework for a sustainable information society. *Journal of Business Ethics* 85: 493–505.

Spiegel-Rösing, I., and D. Price (eds.). 1977. *Science, technology and society: A cross-disciplinary perspective*. London: Sage Publications.

Stirling, A. 2006. Precaution, foresight and sustainability: Reflection and reflexivity in the governance of technology. In *Reflexive governance for sustainable development*, ed. J. Voss and R. Kemp, 225–272. Cheltenham: Edward Elgar.

Stirling, A. 2009. The precautionary principle. In *A companion to the philosophy of technology*, ed. J.-K. Berg Olsen, S.A. Pedersen, and V.F. Hendricks, 248–262. Malden: Wiley-Blackwell.

Tuomela, R. 1991. The social dimension of action theory. *Daimon. Revista de Filosofía* 3: 145–158.

Van den Hoven, J., and J. Weckert (eds.). 2008. *Information technology and moral philosophy*. New York: Cambridge University Press.

Van Gorp, A., and I. Van de Poel. 2008. Deciding on ethical issues in engineering design. In *Philosophy and design: From engineering to architecture*, ed. P. Vermaas et al., 77–89. Dordrecht: Springer.

Verbeek, P.-P. 2008. Morality in design: Design ethics and the morality of technological artifacts. In *Philosophy and design: From engineering to architecture*, ed. P. Vermaas et al., 91–103. Dordrecht: Springer.

Webster, A. 1991. *Science, technology and society*. London: Macmillan.

Weckert, J. 2012. In defense of the precautionary principle. *IEEE Technology and Society Magazine* 31(4): 12–17.

Weckert, J., and J. Moor. 2007. The precautionary principle in nanotechnology. In *Nanoethics: The ethical and social implications of nanotechnology*, ed. F. Allhoff et al., 133–146. Hoboken: Wiley-Interscience.

Weed, D. 2005. Methodologic implications of the precautionary principle: Causal criteria. *Human and Ecological Risk Assessment: An International Journal* 11(1): 107–133.

Wen, H., and D.Y. Yang. 2010. The missing link between technological standards and value-chain governance: The case of patent-distribution strategies in the mobile-communication industry. *Environment and Planning A* 42(9): 2109–2130.

White, L. 1962. The act of invention, causes, context, continuities, and consequences. *Technology and Culture* 3(4): 486–500.

Winner, L. 1977. *Autonomous technology*. Cambridge, MA: The MIT Press.

Winner, L. 1986. *The whale and the reactor: A search for limits in an age of high technology*. Chicago: The University of Chicago Press.

Winner, L. 1993. Upon opening the black box and finding it empty: Social constructivism and the philosophy of technology. *Science as Culture* 16: 427–452. Published as Winner, L. 2003. Social constructivism: Opening the black box and finding it empty. In *Philosophy and Technology: The Technological Condition*, eds. R. C. Scharff and V. Dusek, 233–243. Oxford: Blackwell.

Winner, L. 2003. Luddism as epistemology. In *Philosophy and technology: The technological condition*, ed. R.C. Scharff and V. Dusek, 606–611. Oxford: Blackwell.

World Commission on the Ethics of Scientific Knowledge and Technology. 2005. *Precautionary principle*. Paris: UNESCO.

Zak, P.J. (ed.). 2008. *Moral markets: The critical role of values in the economy*. Princeton: Princeton University Press.

Chapter 2
Values in Engineering and Technology

Ibo van de Poel

2.1 Introduction

There is an intimate relation between technologies and values. Technologies sometimes endanger certain values, like health and safety, as in the case of the Fukushima nuclear disaster. Technologies may also foster certain values, like human well-being, democracy, or privacy. It has even been suggested that technology as such, rather than individual technologies, foster certain values, like efficiency, at the costs of others (e.g., Ellul 1964).

While there has been quite some attention for the relation between values and technology, less attention has been paid to role of values in engineering. I will understand engineering here as an activity that is aimed at understanding, creating, improving, maintaining and dismantling certain technologies. Since technologies are value-laden, it seems natural to expect that values also play, or at least should play, a role in engineering. However, engineering as an activity and as a practice is not only guided by what I will call external values, i.e., values deriving from the social impact of technology, but also by internal values. One might think of such values as technological enthusiasm, which is often a main motive for engineers to develop new technologies, and such values as effectiveness and efficiency, which are largely independent from specific technological applications.

This paper is organized as follows. I start with discussing some of the traditional distinctions that are made in moral philosophy between different kinds of values, especially between instrumental and final value and between intrinsic and extrinsic value. Next, I will discuss and criticize a thesis that is sometimes held with respect to value and technology, i.e., that technology is value-neutral. Thereafter, I will focus on the values in engineering. I will discuss some of the main internal and external values in engineering. I end with conclusions.

I. van de Poel (✉)
Technical University of Delft, TBM-VTI-EFT; Jaffalaan, 5, 2826-BX Delft, The Netherlands
e-mail: I.R.vandePoel@tudelft.nl

© Springer International Publishing Switzerland 2015
W.J. Gonzalez (ed.), *New Perspectives on Technology, Values,
and Ethics*, Boston Studies in the Philosophy and History of Science 315,
DOI 10.1007/978-3-319-21870-0_2

2.2   Final Versus Instrumental and Intrinsic Versus Extrinsic Value

Often a distinction is made between intrinsic and instrumental values. Intrinsic values are those that are good in themselves or for their own sake, while instrumental values are valuable because they help to achieve other values. It should be noted that in this respect an object can be instrumentally valuable and intrinsically valuable at the same time. A car may, for example, be instrumentally valuable as a means of transportation to go from A to B, while at the same time being intrinsically valuable as a beautiful object.

Although the distinction between instrumental and intrinsic value may seem straightforward, it is not. Various philosophers have pointed out a number of terminological and substantive issues with respect to the distinction (for a discussion, see Zimmerman 2004). One issue is that the notion of intrinsic value is ambiguous. The notion is usually understood to refer to objects or states of affairs that are valuable in themselves. Intrinsic value is then value of a non-derivate kind. Intrinsic value may, however, also refer to things that are valuable due to their intrinsic natural, i.e., descriptive, properties. As Christine Korsgaard has pointed out, things that are valuable due to their intrinsic properties are unconditionally good (Korsgaard 1983). Their goodness does not depend on the relation with other objects or with people; otherwise their value would not be intrinsic to the object. However, according to Korsgaard, some things may be good in a non-derivate sense, even if they are not unconditionally good. An example is human happiness in a Kantian respect. According to Kant, human happiness is non-derivate goodness. Happiness is good in itself, and not because it is a means to another end or contributes to another value. Nevertheless, according to Kant, happiness is only conditionally good; it is only good insofar as it corresponds to good will, i.e., respect for the moral law.

To avoid the ambiguity to which Korsgaard refers, I propose to classify the values of objects in two independent ways. The first relates to whether values are relational or not. Values that are not relational will be called "intrinsic values" because these values depend only on intrinsic properties. Otherwise, values are called "extrinsic." The second way relates to whether the values of objects are values for their own sake or not. Values for their own sake will be referred to as 'final values'; otherwise values will be called "instrumental values."

2.3   The Neutrality Thesis

Sometimes the thesis of technology being value-neutral is defended (Florman 1987; Pitt 2000). The main argument usually given for this thesis is that technology is just a neutral means to an end which can be put to good or bad use. Value is thus created during use and is not located in technology. This also means that the objectionable effects of technology are to be blamed on the users and not on technological artifacts, or their designers. As the American Riffle Association has expressed it: "Guns do not kill people, people kill people."

What does claiming that technology is value-neutral exactly entail? One interpretation would be to say that it means that the value of technological artifacts only depends on their extrinsic properties. In this interpretation, the thesis that technology is value-neutral is clearly false. It can be seen as follows. Technological artifacts have a physical or material component, in other words they are also physical objects, even if they are not mere physical objects. The value of physical objects as a means to an end depends – partly at least – on their intrinsic properties. A stone can be used to split a nut thanks to its intrinsic physical properties. A tree leaf would have a much smaller or no instrumental value when it comes to splitting nuts. Since it is implausible that the instrumental value of physical objects merely depends on their extrinsic properties, the same may be said of technologies. So the value of technological artefacts does not only depend on their extrinsic properties.

The thesis that value is not intrinsic to technology may also be interpreted as implying that such value also partly depends on the extrinsic properties of a technology. To judge the plausibility of such a claim, it is crucial to define technology or technological artifacts because to a large extent that is what will determine what we consider to be the intrinsic and extrinsic properties of technological artifacts. If we define technology sufficiently broadly, we can always make values internal to technology. But what happens if we start off with a minimal definition of technology? I think that any plausible minimal account of technology needs to refer to the notion of function, and/or comparable notions like ends, purposes and intentions. The fact technologies have a function implies that they have instrumental value, i.e., that they can be used for some end.

On a minimal definition of technology, then, technology at least has instrumental value. This does not mean that such instrumental value is intrinsic to technological artifacts in the sense that it only depends on the intrinsic properties of technological artifacts. That, indeed, is not usually the case: the particular instrumental value of a particular hammer for driving nails into a piece of wood also depends, for example, on the physical abilities of users and such abilities are extrinsic to the hammer. So even if having instrumental value is part of what it means to be a technical artifact, that same instrumental value is not necessarily intrinsic to the technological artifact.

Van de Poel and Kroes (2014) have argued that technological artefacts cannot only embody instrumental value but also final value. One example they give is a sea dike. The technical function of a sea dike is to prevent the hinterland from flooding, which is instrumental to a moral value like the safety of the inhabitants of the hinterland, which might be considered a final value. The point is not that sea dikes can be used to achieve safety but that achieving safety is part of its *function*. They argue that dikes are *designed for safety*. This is different from, for example, a knife. The function of a knife is cutting; cutting of, for example, bread may be instrumental to a final value like health or survival or human-well-being. However, the attainment of such final values neither is part of the function of knifes nor have normal knifes been designed to achieve such final values. Whereas in the case of the knife, the function of the artifact and the final values that can be achieved by realizing the function are clearly separated this is not the case in the sea dike example. The instrumental function of sea dikes (protection from flooding) can hardly be distinguished from the final value for which they are designed (safety with regard to flooding). After all, the technical function of a dike may be described as providing safety with regard to flooding.

So far we have focused on the value-ladenness of technology; I now want to turn to the value-laden character of engineering. Partly, values in engineering derive from the values realized by technology. Such values are, for example, incorporated in the engineering design process (Van de Poel 2009). Engineering is, however, also value-laden because it is a professional practice (Davis 1998; Pritchard 2009). Michael Davis, for example, has argued that engineering is a profession today. He defines a profession as "a number of individuals in the same occupation voluntarily organized to earn a living by openly serving a certain moral ideal in a morally-permissible way beyond what law, market, and morality would otherwise require" (Davis 1998, p. 417). Engineering as a profession is, in his view, thus by definition value-laden.

The statement that engineering is value-laden is not uncontroversial. Samuel Florman, has, for example suggested that it is not the task of engineers to determine the broader social goals for which technology is to be used or to be optimized (Florman 1983). Somewhat similarly, Steven L. Goldman talks about the social captivity of engineering. According to him engineering practice is "captive to social determinants of technological action that selectively exploit engineering expertise, define the problems engineers are to address as the terms of acceptable solution …" (Goldman 1991, p. 121).

Heinz Luegenbiehl has explicitly addressed the question whether a definition of engineering should "emphasize the requirement of engineering activity to benefit humanity" (Luegenbiehl 2010, p. 153) or should choose a more value-neutral approach. He opts for the latter option and defines engineering as "the transformation of the natural world, using scientific principles and mathematics, in order to achieve some desired practical end" (Luegenbiehl 2010, p. 153). He maintains nevertheless that "some value element is unavoidable, in that I assume that engineering activity should leave the world no less well off and that disbenefits created by engineering not be catastrophic in nature" (Luegenbiehl 2010, p. 153).

In what follows, my aim is to further explore the values that play a role in engineering. In doing so, I will distinguish between what I will call internal and external values.

Internal values are values that are perceived by engineers as internal to engineering practice and that do not, or at least seemingly do not, refer to broader social goals and values. Internal values are typically context-independent, in the sense that they are relevant in various contexts of use. A typical example is efficiency; efficiency is an important value in engineering independent from the exact technology or the exact context of usage. Similarly, a value like technological enthusiasm is more or less independent from the technology developed. Internal values are often, although not necessarily always, perceived as final by engineers, i.e., as values that are strived for their own sake. However, as we will see below from a moral point of view internal values are usually not final values.

External values are values that are related to effects of technology on other practices. Typical examples are safety, health and sustainability. They typically refer to broader human, social, environmental, and political goals. External values may be final in a moral sense, and they often are as we will see, but this is not necessarily the case. Although external values find their origin outside engineering practice, they may be internalized, for example through technical codes and standards. This has typically happened with a value as safety, as will see in more detail below and is increasingly happening with sustainability.

2.4 Internal Values

2.4.1 Technological Enthusiasm

Technological enthusiasm pertains to the ideal of wanting to develop new technological possibilities and take up technological challenges. This is an ideal that motivates many engineers. It is fitting that Samuel Florman (1994/1976) refers to this as "the existential pleasures of engineering." One good example of technological enthusiasm is the development of Google Earth, a programme with which, via the Internet, it is possible to zoom in on the earth's surface. It is a beautiful concept but it gives rise to all kinds of moral questions, for instance in the area of privacy (you can study the opposite neighbour's garden in great detail) and in the field of security (terrorists could use it to plan attacks). In a recent documentary on the subject of Google Earth one of the programme developers admitted that these are important questions.[1] Nevertheless, when developing the programme these were matters that the developers had failed to consider because they were so driven by the challenge of making it technologically possible for everyone to be able to study the earth from behind his or her PC.

Technological enthusiasm in itself is not morally improper; it is in fact positive for engineers to be intrinsically motivated as far as their work is concerned. The inherent danger of technological enthusiasm lies in the possible negative effects of technology and the relevant social constraints being easily overlooked. This has been exemplified by the Google Earth example. It is exemplified to an extreme extent by the example of Wernher von Braun.

Wernher von Braun is famous for being the creator of the space programme that made it possible to put the first person on the moon on 20th July 1969. von Braun grew up in Germany. From an early age he was fascinated by rocket technology. In the 1930s von Braun was involved in developing rockets for the German army. In 1937 he joined Hitler's National Socialist Party and in 1940 he became a member of

---

[1] "Google: Achter het scherm" (i.e. "Google: Behind the Screen"), *Tegenlicht*, broadcast on May 7, 2006.

the SS. There is much to indicate that von Braun's main reason for wanting to join
the SS was carefully calculated: in that way he would be able to continue his important work in the field of rocket technology. During the Second World War von Braun
played a major role in the development of the V2-rocket which was deployed from
1944 onwards to bomb, amongst other targets, the city of London. When, in 1945,
von Braun realised that the Germans were going to lose the war he arranged for his
team to be handed over to the Americans. In the United States von Braun originally
worked on the development of rockets for military purposes but later he fulfilled a
key role in the space travel programme, a programme that was ultimately to culminate
in man's first steps on the moon. Von Braun's big dream did therefore ultimately
come true.

Von Braun was reconciled to the subordinate role of engineers but perpetually
sought ways of pursuing his technological ideals and, in so doing, displayed a
degree of indifference to the social consequences of the application of his work and
to the immoral intentions of those who had commissioned the task. His creed must
have been: "In times of war, a man has to stand up for his country, as a combat
soldier as a scientist or as an engineer, regardless of whether or not he agrees with
the policy his government is pursuing" (Stuhlinger and Ordway 1994, p. xiii). It is
a role that might alternatively be described as being that of a "hired gun." The
dangerous side of this role can perhaps best be summed up in the words of the song
text of the British satirist Tom Lehrer[2]:

> Once the rockets go up
> Who cares where they come down
> 'that's not my department'
> said Wernher von Braun.

2.4.2   Effectiveness and Efficiency

Engineers tend to strive for effectiveness and efficiency. Effectiveness can be defined
as the degree to which an artifact fulfils its function. Efficiency could be defined as
the ratio between the degree to which an artifact fulfils its function and the effort
required to achieve that effect. Efficiency in the modern sense is usually construed
as an output/input ratio (Alexander 2009). The energetic efficiency of a coal plant
may thus be defined as the ratio between the energy contained in the power produced and the thermal energy contained in the unburnt coal.

Effectiveness and efficiency are different values that may well conflict. The
design that most effectively fulfils its intended function may not necessarily be the
most efficient one. A very effective vacuum cleaner that removes more dust than a
less effective one may nevertheless be less energy-efficient, that is to say, it may use

---

[2] Text from the number "Wernher von Braun" by Tom Lehrer that featured in his album *That was
the year that was* of 1965.

more energy per unit of dust removed than the less effective vacuum cleaner. So, we may be faced with a conflict between effectiveness and efficiency.

The drive to strive towards effectiveness and efficiency is an attractive value for engineers because it is – apparently – so neutral and objective. It does not seem to involve any political or moral choices, which is something that many engineers experience as subjective and therefore wish to avoid. Efficiency is also something that in contrast, for example, to human welfare can be defined by engineers and is also often quantifiable. Engineers are, for example, able to define the efficiency of the energy production in an electrical power station and they can also measure and compare that efficiency.

Efficiency is an ideal that endows engineers with authority because it is something that – at least at first sight – one can hardly oppose and that can seemingly be measured objectively. From a moral point of view, however, effectiveness and efficiency are not always worth pursuing. That is because effectiveness and efficiency suppose an external goal in relation to which they are measured. That external goal can be to consume a minimum amount of non-renewable natural resources to generate energy, but also war or even genocide. It was no coincidence that Nazi bureaucrats like Eichmann were proud of the efficient way in which they were able to contribute to the so-called 'resolving of the Jewish question' in Europe which was to lead to the murdering of six million Jews and other groups that were considered inferior by the Nazis like Gypsies and mental patients (Arendt 1965). The matter of whether effectiveness or efficiency is morally worth pursuing therefore depends very much on the ends for which they are employed. So, although some engineers have maintained the opposite, the measurement of the effectiveness and efficiency of a technology is value-laden. It proposes a certain goal for which the technology is to be employed and that goal can be value-laden. Moreover, to measure efficiency one need to calculate the ratio between the output (the external goal) and the input, and also the choice of the input may be value-laden. A technology may for example be efficient in terms of costs but not in terms of energy consumption.

2.4.3 Other Internal Engineering Values

There are a range of other internal values to engineering. I mention some:

• *Reliability*, which might be understood as "the ability of a product to perform its function adequately over a period of time without failing" (cf. Kuo et al. 2001, p. 252).

• *Robustness,* which may be defined as the "ability of a product to perform its function adequately in new or unforeseen circumstances" (cf. Vermaas et al. 2011, p. 113).

• *Maintainability* which might be understood as "the probability that a failed system can be repaired in a specific interval of downtime against reasonable cost" (cf. Kuo et al. 2001, p. 251).

- *Compatibility* which might be understood as "the ability of a product to adequately perform its function in conjunction with other apparatus and infrastructure."
- *Quality*. Quality might be understood in a variety of ways. Sometimes it is used to refer to such values as reliability, robustness and compatibility. It is also used in the sense of "robust in meeting the requirements (within certain acceptable limits) despite variations in the production process" (cf. Holt and Barnes 2010, p. 125). It might also be understood in terms of "meeting or even exceeding user requirements" or in terms of "user satisfaction."[3] In the latter case, it seems to refer to an external value because user requirements and user satisfaction refer to values outside engineering practice.
- *Rationality*. Rationally does not so seem to refer to a value that is realized in the products developed in engineering but rather to engineering as a process. It relates to how this process is organized, how decisions are made and how knowledge is developed. Rationality in engineering can be understood in a range of different ways; for a good discussion see Kroes et al. (2009).

Most of these values are internal values in the sense that engineers value them independent from the exact technology they develop and independent from particular applications. While engineers may perceive these values as final, just like they value technological enthusiasm and effectiveness and efficiency as final, from a moral point of view they are instrumental values, with the possible exception of rationality.

A number of approaches have been developed to design for the mentioned internal values. Such approaches are now known under the heading: design for X or DFX (Holt and Barnes 2010; Kuo et al. 2001). In DFX approaches, X can stand for a certain virtue or value or for a life phase. Table 2.1 lists a number of DFX$_{virtue}$ and DFX$_{lifephase}$ approaches that are distinguished in a recent overview article by Holt and Barnes (2010).

2.5   External Values

2.5.1   Safety and Health

Safety and health are without doubt among the main external values in engineering. Most US codes of ethics declare these values to be paramount in engineering. So, the NSPE Code of conduct states that "Engineers shall hold paramount the safety, health, and welfare of the public." Likewise, the code of ethics of the FEANI, the overarching European association of engineering societies, states that "Engineers

---

[3] If it is used in terms of "user satisfaction," quality seems to refer to the value of human well-being of a desire-satisfaction account of well-being is adopted. Cf. the discussion below.

Table 2.1 DFX approaches (Holt and Barnes 2010)

DFX$_{virtue}$	DFX$_{lifephase}$
Design for environment	Design for manufacture and assembly
Design for quality	Design for end-of-life
Design for maintainability	Design for disassembly
Design for reliability	Design for recycling
Design for cost	Design for supply chain
Affective design	
Inclusive design	

shall carry out their tasks so as to prevent avoidable danger to health and safety, and prevent avoidable adverse impact on the environment."

Safety is sometimes defined as the absence of risk and hazards. However, risk reduction is not always feasible or desirable. It is sometimes not feasible, because there are no absolutely safe products and technologies. But even if risk reduction is feasible it may not be desirable from a moral point of view. Reducing risk often comes at a cost. Safer products may be more difficult to use, more expensive or less sustainable. So sooner or later, one is confronted with the question: what is safe enough? What makes a risk (un)acceptable? The ethical literature on risk has established that the moral acceptability of risks does not only depend on their magnitude but also on considerations like voluntariness, the balance and distribution of benefits and risks, and the availability of alternatives (Asveld and Roeser 2009; Hansson 2003; Shrader-Frechette 1991; Hansson 2009; Harris et al. 2008). So conceived, safety refers to the situation in which the risks have been reduced in as far that is reasonably feasible and desirable.

Health is defined by the World Health Organisation (WHO) as "state of complete physical, mental and social well-being and not merely the absence of disease or infirmity" (World Health Organization 2006). This definition refers to the broader value of human well-being that I will discuss below. In engineering, the focus is usually on avoiding negative influences on human health. It is not obvious that there is a requirement for engineering to contribute positively to human health, with the exception perhaps of some specific domains like health technologies. The possibilities of new technologies, like biotechnology and nanotechnology, have also led to a debate on whether technology should only aim at curing illness and perhaps improving health or should also contribute to improving humans and their achievements (Savulescu and Bostrom 2009). The latter is known as human enhancement. The positions on human enhancement range from the belief that it is not only desirable but even morally required to the conviction that it is utterly undesirable and immoral.

Health and safety are often seen as final values from a moral point of view. It might also be argued that these values are not really valuable in themselves bur rather contribute to the good life or human well-being. Their contribution is, however, not merely causal, but rather they are, as values, constitutive for the overarching value of human well-being. Safety and health may thus also be seen as constitutive values for the final value of human well-being.

In regular engineering practice, however, the focus is usually on avoiding negatively influencing human health. In such cases, the focus is often on potential health risks that are to be minimized. The approach may then be similar to that of safety risks that I discussed above with an important role for the notion of acceptable risks. For example, for potential toxic substances, acceptable health risks are often formulated in terms of acceptable daily intake (Covello and Merkhofer 1993).

Although health and safety are external values, in the sense that they refer to the effects of technology outside engineering practice, they have been internalised in engineering practice over time. In the case of safety this has even led to the academic treatment of the principles of safety engineering, although safety engineering is fragmented over different technological areas (Hansson 2009). Health has not yet led to a specific area in engineering, although in the late nineteenth and in the twentieth century attempts have been made to establish sanitary engineering and later public health engineering as distinct disciplines (Van de Poel 2008, pp. 614–615, footnote 9).

Safety and health are also internalized in engineering through technical codes and standards. Technical codes are legal requirements that are enforced by a governmental body to protect safety, health and other relevant values. Technical standards are usually recommendations rather than legal requirements that are written by engineering experts in standardization committees. Codes and standards have two main functions (Hunter 1997). The first is standardization and the promotion of compatibility. The second aim of codes and standards is guaranteeing a certain quality or protecting external values. Though external values usually are not explicitly stated in codes and standards, considerations in safety and health often are the foundation for the content of codes and standards.

2.5.2 Human Well-Being

Several engineering codes of ethics state that "engineers shall use their knowledge and skill for the enhancement of human welfare" (Code of ethics American Society of Civil Engineers and Code of Ethics American Society for Mechanical Engineering.) Also in other engineering texts and methods, one find references to external values like human welfare, happiness, quality of life, human flourishing, the good life, and well-being. I will here use the term "human well-being" to refer to the value that is at stake in all these cases. I take it that well-being not only refers to feeling well here and now but that it tells something about how somebody's life is going *for* that person.

In moral philosophy, human well-being is generally seen as a final value, that is worthwhile for its own sake, rather than to achieve something else. In philosophy, three main theories about how to understand the value of well-being have been developed (Crisp 2008):

- Hedonism conceives of human well-being as pleasurable experience.
- Desire satisfaction accounts conceives well-being as the fulfillment of the (actual) desires that people have
- Objective list accounts assume that well-being can be understood in terms of a list of general prudential values

I will first briefly discuss some of the philosophical objections that have been raised in philosophy against each of these theories and will then discuss how each of these philosophical positions may be put to work in the context of engineering, particularly in engineering design.

A main objection against experience accounts has been raised by Nozick (1974), who invites us to imagine an *Experience Machine* that can give us any possible, positive experience we desire, while we actually float in tank and do nothing. Would you plug in to this *Experience Machine*? While most of us would probably appreciate the pleasure and joy that the machine can create for us, it seems likely that many people would not plug in. The reason is that what we not just value experiences but also value *to be* somebody and *to do* certain things. We do not just want the experience of friendship, we want friendship; we do not just want the impression of being in control of our own life but we want to be in control (at least to some extent). We can thus conclude that what make positive experiences good or desirable are the *values* on which they are based. Sometimes the value lies in the experience itself (as with the value of joy and pleasure); in other cases the values lies outside the experience itself (as with values like accomplishment, friendship and autonomy).

Desire-satisfaction accounts also have a number of problems (Crisp 2008; Griffin 1986). One problem is that people might well desire things that do not contribute to their well-being. I may have a longing desire to eat an entire pie every day, but on closer reflection, it is likely that I will come to the conclusion that that is not contributing to my well-being. Well-being then is not so much about satisfying the desires I have here and now but rather about how my life overall and over a longer period of time is going. Another problem of desire-satisfaction accounts is a phenomenon known as adaptive preferences (Nussbaum 2000, pp. 136–142). When people are for a long time deprived of basic rights or needs, they might very well loose a desire for such rights or for fulfilling those needs. It would however be wrong to conclude that fulfilling those rights and needs then no longer contributes to their well-being. In fact most people would start appreciating those rights and the fulfillment of those needs again once they are no longer deprived of them.

Objective list accounts assume that it is possible to list a number of values (or other items such as capabilities) that together constitute well-being. Objective list accounts also have their problems. First, it seems rather obscure how we can come to a list of objective prudential values and how we know when it is complete. Second, such accounts seem to ignore reasonable differences between people in what constitutes well-being for them. After all, my well-being is to an important extent dependent on my ability to set my own goals in life and to accomplish these (Raz 1986). One possible way to try to avoid these problems is by basing the list not only on certain features of human nature but also in part on so-called informed desires (see Griffin 1986, p. 70; Nussbaum 2000, p. 76). These are basically the desires that people were to have if they were fully (or at least sufficiently) informed and took a reflective attitude towards their own life. In addition it could be argued that even if the resulting list would consist in the basic components of human-well-being, that the specific content these abstract values get in the life of individual people and their relative importance may well reasonably differ from person to person (and maybe also between cultures).

The three philosophical approaches to well-being can also be found in engineering, particularly in engineering design. Several authors have argued that user value, and in particular human well-being, is created through experience and they have developed approaches to measure the experiences created by technical products and to design for certain experiences (e.g. Koskinen et al. 2003; Desmet and Hekkert 2007). Desire-satisfaction accounts are perhaps the most influential in engineering, as they fit well with economic theory and therefore with approaches that focus on adding economic value. An example is demand modeling (Cook and Wu 2001). Also approaches that focus on quality management, in particular quality function deployment (QFD) are based on a desire-satisfaction account of well-being. QFD aims at systematically taking into account user satisfaction in the engineering process by systematically translating customer demands in engineering characteristics and setting priorities amongst them (Akao 1990; Hauser and Clausing 1988). Although the method is beset with some methodological problems (Van de Poel 2007), it is a main example of how well-being conceived in desire-satisfaction terms can be taken into account in engineering.

Objective list accounts have until yet not been very influential in engineering and design. However, a number of authors have sketched how an approach based on an objective list account of well-being may guide engineering design. Van de Poel (2012) provides a general discussion on how we might design for well-being if we adopt an objective list account. Another approach may to understand the values in an objective list account in terms of human capabilities, an approach that has been especially advocated by Sen and by Nussbaum (Sen 1985; Nussbaum 2000). They developed the capability approach as an alternative to economic approaches to well-being. Oosterlaken (2009) gives some ideas about how one might design for capabilities.

2.5.3 Sustainability

Although environmental values play a role in engineering for quite some time, the last decade this has been increasingly understood in terms of the broader value of sustainability. Thus the Code of conduct of the US NSPE (National Society of Professional Engineers) states that "Engineers are encouraged to adhere to the principles of sustainable development in order to protect the environment for future generations." It is interesting that this is formulated in terms of a recommendation rather than a requirement to hold paramount as in the case of safety, health and human well-being. This suggests that sustainability is still a less generally accepted value in engineering that the aforementioned ones, although this may now be changing. As we will see below, this may be partly due to the contested character of sustainability.

The most influential definition of sustainable development has been provided by the Brundlandt commission:

"Sustainable development is development that meets the needs of the present without compromising the ability of future generations to meet their own needs. It contains within it two key concepts:

• the concept of 'needs', in particular the essential needs of the world's poor, to which over-riding priority should be given; and

• the idea of limitations imposed by the state of technology and social organization on the environment's ability to meet present and future needs" (WCED 1987).

It can be argued that this definition of sustainable development refers to two types of justice, i.e., intergenerational justice, as testified in the phrase "without compromising the ability of future generations to meet their own needs" and intragenerational justice, as testified in the phrase "the essential needs of the world's poor, to which over-riding priority should be given" (Brumsen 2011). It should be noted that both types of justice might conflict in particular cases. A typical example is biofuels. Biofuels are attractive in terms of intergenerational justice because they make energy resources available to future generations and they may help to abate greenhouse warming. At the same time, they compete with food crops, so contributing to increasing food prices, which may lead to an increase in malnutrition and hunger, especially amongst the world poor. From the viewpoint of intragenerational justice, biofuels therefore do not (yet) seem very attractive.

Another issue with respect to sustainability is whether should be understood in anthropocentric or in biocentric terms (Brumsen 2011). If sustainability is understood in anthropocentric terms, sustaining nature and the environment are strived for the sake of human well-being; on a bio-centric view, nature is attributed final value, i.e., value independent from human goals and human well-being. It appears to me that a plausible understanding of sustainability should somehow take into account both the sustenance of human well-being, and nature as finally valuable.

It might then be argued that sustainability is an overarching value that refers to, at least, four constitutive values: intragenerational justice, intergenerational justice, human well-being and its sustenance and nature as a final value and its sustenance. Even if one accepts that all these values are somehow constitutive for sustainability, one might disagree about the exact understanding of each of these values, as we already saw for human well-being above, and about their relative importance. Sustainability might then be understood as a "contested value" (Jacobs 1999).

Typical for contested concepts, according to Jacobs (1999), is that while there is agreement that they are valuable there is disagreement about their exact meaning and content. Engineers often seem to dislike the contested character of sustainability; it also seems to have been a reason for some engineering societies not to include it in their codes of ethics or to make it only a recommended rather than a required value as we saw in the case of the NSPE code.

One might try to overcome the contested character of sustainability, by trying to reach consensus on a generally accepted definition of sustainability. Such an approach seems me, however, illusionary as disagreements about sustainability are often disagreements about the kind of society we want to live in, and such disagreements are ineradicable in the pluralist society we live in. This does not mean, however, that we cannot take into account sustainability in engineering and design. Often engineering solutions will mainly be aimed at taking away existing unsustainability or avoiding adding new unsustainability. In many cases, agreement about what is unsustainable is much easier to achieve than agreement about what is sustainable.

As a value, sustainability is increasingly internalized in engineering practice in a number of ways. First, it plays a role in engineering trough laws and regulations, and through technical codes and standards. One might for example think of requirements for energy efficiency of devices, or requirements for heat isolation. There is also an increasing attention for what might be called design for sustainability (Bhamra and Lofthouse 2007; Birkeland 2002). Such approaches may state general design principles for sustainability, provide tools to design for sustainability, and suggest certain technical features or design concepts. There are also an increasing number of tools for sustainability in engineering, one might in particular think of various tools for life cycle analysis of products.

2.5.4   Other External Values

In addition to the aforementioned values, other external values are relevant for engineering. Some of these external values are generally relevant for engineering. Examples are justice and democracy, and inclusiveness. For such values, also approaches have been developed to give them a larger role in engineering practice. Sclove (1995) for example has formulated design principles for democratic technologies. For inclusive design, a whole range of approaches has been developed, that aim at making accessible technological products that all users, with special attention for the underprivileged, like for example handicapped people (Clarkson 2003; Erlandson 2008).

In addition to such more general external values, one might distinguish external values that are more domain-specific. A typical example is aesthetics in architecture. Friedman et al. (2006) have distinguished 12 values that are especially important in the domain of information and communication technologies (ICTs): human welfare, ownership and property, privacy, freedom from bias, universal usability, trust, autonomy, informed consent, accountability, identity, calmness and environmental sustainability.

She and her colleagues have also developed an approach for integrating such values in design: value sensitive design (VSD). Friedman and Kahn (2003) distinguish three kinds of investigations that are relevant to VSD: empirical, conceptual and technical. Empirical investigations "involve social scientific research on the understanding, contexts, and experiences of the people affected by technological designs" (Friedman and Kahn 2003, p. 1187). It is not hard to see why this is relevant: people's experiences, contexts and understanding are certainly important when it comes to appreciating precisely what values are at stake and how these values are affected by different designs. Conceptual investigations aim at clarifying the values at stake, and at making trade-offs between the various values. Technical investigations "involve analyzing current technical mechanisms and designs to assess how well they support particular values, and, conversely, identifying values, and then identifying and/or developing technical mechanisms and designs that can support those values" (Friedman and Kahn 2003, p. 1187). The second part of this assertion is especially interesting and relevant because it provides the opportunity to develop new technical options that more adequately meet the values of ethical importance than do current options.

2.6   Conclusions

My main aim in this contribution was to explore some of the main internal and external values in engineering. The treatment of these values has necessarily been somewhat cursory. Nevertheless, I think that the overview given contains some general lessons with respect to internal and external values in engineering and their relation. Internal values like technological enthusiasm and efficiency are often perceived by engineers as final. However, in a moral sense, they are usually instrumental values; they are means to achieve final values that are usually external to engineering practice. This is not to say that internal values are morally improper, but that their moral appropriateness depends on the broader, final values for which they are put to work for.

Most of the external values I discussed are final values or they are constitutive values, i.e., values that are constitutive for some final value, for example by being a part of the overarching final value, rather than by just being a means to the final value. External values seem relevant for engineering practice in at least two ways. First, they may provide part of the explanation and justification why certain internal values like efficiency are strived for in particular engineering projects. Second they may be more directly relevant in engineering practice. As we have seen they may be internalized, for example through technical codes and standards or specific engineering approaches, like Quality Function Deployment or Design for X approaches.

It is important to be aware that if external values are to play out in engineering they have to be internalized, at least to some extent, in engineering. Obviously, this process of internalization has been taken place in most engineering domains with respect to the values of safety and health; it is now also increasingly occurring for human well-being and sustainability. Domain specific values like aesthetics and privacy have also to a large extend been internalized in the relevant domains. With respect to other values and other domains, this process of internalization is often just starting.

Acknowledgements Section 2.2 contains two paragraphs and Sect. 2.3 contains three paragraphs from: Van de Poel 2009. "Values in Engineering Design." In A. Meijers (ed.), *Handbook of the Philosophy of Science. Volume 9: Philosophy of Technology and Engineering Sciences.* Oxford: Elsevier, pp. 973–1006. Section 2.4.1 and 2.4.2 contain some paragraphs from van de Poel, I. and Royakkers, L. 2011. *Ethics, Technology and Engineering.* Oxford: Wiley-Blackwell, (sections 1.4.1 and 1.4.1 in the book). Section 2.5.2 contains two paragraphs from Van de Poel 2012. "Can We Design for Well-Being?" In P. Brey, A. Briggle and E. Spence (eds.), *The Good Life in a Technological Age.* N. York: Routledge, pp. 295–306. Some parts of this text have also appeared in German as van de Poel, I. 2013. "Werthaltigkeit Der Technik." In A. Grundwald (ed.), *Handbuch Technikethik.* Stuttgart/Weimar: Verlag J. B. Metzler, pp. 133–137.

This paper was written as part of the research program 'New Technologies as Social Experiments', which is supported by the Netherlands Organization for Scientific Research (NWO) under grant number 277-20-003.

References

Akao, Y. (ed.). 1990. *Quality function deployment. Integrating customer requirements into product design.* Cambridge, MA: Productivity Press.

Alexander, J.K. 2009. The concept of efficiency: An historical analysis. In *Handbook of the philosophy of science. Volume 9: Philosophy of technology and engineering sciences,* ed. A. Meijers, 1007–1030. Oxford: Elsevier.

Arendt, H. 1965. *Eichmann in Jerusalem: A report on the banality of evil.* New York: Penguin Classics.

Asveld, L., and S. Roeser (eds.). 2009. *The ethics of technological risk.* London: Earthscan.

Bhamra, T., and V. Lofthouse. 2007. *Design for sustainability: A practical approach.* Aldershot: Gower.

Birkeland, J. 2002. *Design for sustainability. A source book for ecological integrated solutions.* London: Earthscan.

Brumsen, M. 2011. Sustainability, ethics and technology. In *Ethics, technology, and engineering,* ed. I. van de Poel and L. Royakkers, 277–300. Oxford: Wiley-Blackwell.

Clarkson, J. 2003. *Inclusive design: Design for the whole population.* London/New York: Springer.

Cook, H.E., and A. Wu. 2001. On the valuation of goods and selection of the best design alternative. *Research in Engineering Design* 13: 42–54.

Covello, V.T., and M.W. Merkhofer. 1993. *Risk assessment methods. Approaches for assessing health and environmental risks.* New York/London: Plenum.

Crisp, R. 2008. Well-being. In *The Stanford encyclopedia of philosophy (Winter 2008 Edition),* ed. E.N. Zalta, http://plato.stanford.edu/archives/win2008/entries/well-being/. Accessed on 22 Oct 2014.

Davis, M. 1998. *Thinking like an engineer. Studies in the ethics of a profession.* New York/Oxford: Oxford University Press.

Desmet, P., and P. Hekkert. 2007. Framework of product experience. *International Journal of Design* 1(1): 57–66.

Ellul, J. 1964. *The technological society.* New York: Knopf.

Erlandson, R.F. 2008. *Universal and accessible design for products, services, and processes.* Boca Raton: CRC Press.

Florman, S.C. 1983. Moral blueprints. In *Engineering professionalism and ethics,* ed. H. Schaub, K. Pavlovic, and M.D. Morris, 76–81. New York: John Wiley and Sons.

Florman, S.C. 1987. *The civilized engineer.* New York: St. Martin's Press.

Florman, S.C. 1994/1976. *The existential pleasures of engineering.* New York: St. Martin's Press.

Friedman, B., and P.H.J. Kahn. 2003. Human values, ethics and design. In *Handbook of human-computer interaction,* ed. J. Jacko and A. Sears, 1177–1201. Mahwah: Lawrence Erlbaum Associates.

Friedman, B., P.H.J. Kahn, and A. Borning. 2006. Value sensitive design and information systems. In *Human-computer interaction in management information systems: Foundations*, ed. P. Zhang and D. Galletta, 348–372. Armonk: M. E. Sharpe.

Goldman, S.L. 1991. The social captivity of engineering. In *Critical perspectives on non-academic science and engineering*, ed. P.T. Durbin, 121–145. Bethlehem: Lehigh University Press.

Griffin, J. 1986. *Well-being: Its meaning, measurement, and moral importance*. Oxford [Oxfordshire]/New York: Clarendon.

Hansson, S.O. 2003. Ethical criteria of risk acceptance. *Erkenntnis* 59: 291–309.

Hansson, S.O. 2009. Risk and safety in technology. In *Handbook of the philosophy of science. Volume 9: Philosophy of technology and engineering sciences*, ed. A. Meijers, 1069–1102. Oxford: Elsevier.

Harris, C.E., M.S. Pritchard, and M.J. Rabins. 2008. *Engineering ethics. Concepts and cases*. Belmont: Wadsworth.

Hauser, J.R., and D. Clausing. 1988. The house of quality. *Harvard Business Review* 66(3): 63–73.

Holt, R., and C. Barnes. 2010. Towards an integrated approach to 'Design for X': An agenda for decision-based DFX research. *Research in Engineering Design* 21(2): 123–136.

Hunter, T.A. 1997. Designing to codes and standards. In *Asm handbook*, ed. G.E. Dieter and S. Lampman, 66–71. Materials Park: ASM.

Jacobs, M. 1999. Sustainable development as contested concept. In *Fairness and futurity. Essays on environmental sustainability and social justice*, ed. A. Dobson, 21–45. Oxford: Oxford University Press.

Korsgaard, C.M. 1983. Two distinctions in goodness. *Philosophical Review* 92(2): 169–195.

Koskinen, I., K. Battarbee, and T. Mattelmäki (eds.). 2003. *Emphatic design. User experience in product design*. Helsinki: IT Press.

Kroes, P., M. Franssen, and L. Bucciarelli. 2009. Rationality in design. In *Handbook of the philosophy of science. Volume 9: Philosophy of technology and engineering sciences*, ed. A. Meijers, 565–600. Oxford: Elsevier.

Kuo, T.-C., S.H. Huang, and H.-C. Zhang. 2001. Design for manufacture and design for 'X': Concepts, applications, and perspectives. *Computers & Industrial Engineering* 41(3): 241–260.

Luegenbiehl, H.C. 2010. Ethical principles for engineers in a global environment. In *Philosophy and engineering. An emerging agenda*, ed. I. Van de Poel and D.E. Goldberg, 147–160. Dordrecht: Springer.

Nozick, R. 1974. *Anarchy, state, and utopia*. New York: Basic books.

Nussbaum, M.C. 2000. *Women and human development. The capabilities approach*. Cambridge: Cambridge University Press.

Oosterlaken, I. 2009. Design for development: A capability approach. *Design Issues* 25(4): 91–102.

Pitt, J.C. 2000. *Thinking about technology. Foundations of the philosophy of technology*. New York: Seven Bridges Press.

Pritchard, M.S. 2009. Professional standards in engineering practice. In *Handbook of the philosophy of science. Volume 9: Philosophy of technology and engineering sciences*, ed. A. Meijers, 953–971. Oxford: Elsevier.

Raz, J. 1986. *The morality of freedom*. Oxford: Oxford University Press.

Savulescu, J., and N. Bostrom. 2009. *Human enhancement*. Oxford: Oxford University Press.

Sclove, R.E. 1995. *Democracy and technology*. New York: The Guilford Press.

Sen, A. 1985. *Commodities and capabilities*. Amsterdam/New York: Elsevier.

Shrader-Frechette, K.S. 1991. *Risk and rationality. Philosophical foundations for populist reform*. Berkeley: University of California Press.

Stuhlinger, E., and F.I. Ordway III. 1994. *Wernher von Braun, crusader for space: A biographical memoir*. Malabar: Krieger.

Van de Poel, I. 2007. Methodological problems in QFD and directions for future development. *Research in Engineering Design* 18(1): 21–36.

Van de Poel, I. 2008. The bugs eat the waste: What else is there to know? Changing professional hegemony in the design of sewage treatment plants. *Social Studies of Science* 38(4): 605–634.

Van de Poel, I. 2009. Values in engineering design. In *Handbook of the philosophy of science. Volume 9: Philosophy of technology and engineering sciences*, ed. A. Meijers, 973–1006. Oxford: Elsevier.

Van de Poel, I. 2012. Can we design for well-being? In *The good life in a technological age*, ed. P. Brey, A. Briggle, and E. Spence, 295–306. New York: Routledge.

Van de Poel, I., and P. Kroes. 2014. Can technology embody values? In *Moral agency and technical artefacts*, ed. P. Kroes and P.-P. Verbeek, 103–124. Dordrecht: Springer.

Vermaas, P., P. Kroes, I. van de Poel, M. Franssen, and W. Houkes. 2011. A philosophy of technology: From technical artefacts to sociotechnical systems. *Synthesis Lectures on Engineers, Technology and Society* 6(1): 1–134.

WCED. 1987. *Our common future. Report of the World Commission on Environment and Development*. Oxford: Oxford University Press.

World Health Organization. 2006. Constitution of the World Health Organization ‒ Basic documents. Supplement.

Zimmerman, M.J. 2004. Intrinsic Vs. Extrinsic value. In *The Stanford encyclopedia of philosophy (Fall 2004 Edition)*, ed. E.N. Zalta, http://plato.stanford.edu/archives/fall2004/entries/value-intrinsic-extrinsic/. Accessed on 22 Oct 2014.

Chapter 3
Values Regarding Results of the Information and Communication Technologies: Internal Values

Paula Neira

Information and communication technologies (ICTs), through a setting-up of specific aims—and, therefore, the processes that they carry out in order to achieve them—commonly produce a wide variety of repercussions. When the technology is designed in an efficient way, the results can be linked with the objectives previously selected.[1] But there can also be technologies of a different kind: those whose aims are now different from the objectives for which were initially conceived. Certainly, this is true in the case of the Internet.

Besides the evaluation of the aims and processes, technology requires that its results be evaluated as well. This task is carried out according to values, first from an internal point of view and later from an external perspective. Thus, the evaluation of the Internet as a technological product can be considered in two different ways: on the one hand, from the internal point of view, through the analysis of the adaptation of the mediums for the prefixed aims; and on the other hand, from an external perspective that evaluates the adaptation of the technology mentioned within the context of society.

Regarding the internal aspect of the technological product, the degree of technological accessibility, its versatility, its level of internal profitability or efficacy, and obviously its efficiency should all be borne in mind. Seen from the angle of the aims, processes and results, there are some relationships with the environment (social, cultural, political, etc.) in ICT. It is necessary to evaluate other aspects that affect the external dimension of technology: the questions related to aesthetics,

---

[1]An analysis of some aspects of technological design can be found in Oosterlaken and Hoven (2012).

P. Neira (✉)
CSHG, Universidad de Santiago de Compostela, Ctra./ Santiago-Noia, 1,
15896-A Barcia, A Coruña, Spain
e-mail: pneira@cshg.es

© Springer International Publishing Switzerland 2015   47
W.J. Gonzalez (ed.), *New Perspectives on Technology, Values,*
and Ethics, Boston Studies in the Philosophy and History of Science 315,
DOI 10.1007/978-3-319-21870-0_3

ergonomics, ecology, ethics, social and political consequences, and the economic repercussion.

Economic repercussion can also be reviewed from an internal dimension, insofar as the economic values have results that are at internal as well as at external level (cf. Gonzalez 1999). In order to make an evaluation of Internet as a technological product, there are two main views: (a) its qualitative innovation, which emphasizes what has not been done previously; and (b) its quantitative innovation, which checks the ability to achieve someone's objectives more effectively.

About the internal evaluation, Simon considers that "fulfillment of purpose or adaptation to a goal involves a relation among three terms: the purpose or goal, the character of the artifact, and the environment in which the artifact performs" (Simon 1996, p. 5). To clarify this relationship he gives the example of a clock. The clock is intended to measure time, so its structure has to be designed for that purpose, but the environment can affect the performance of the task, as a solar clock understandably does not work in Alaska. Therefore, that an artifact works and meets the objectives for which it has been designed depends not only on its internal structure but also on the environment in which it is located.

In this sense, the targets and the environment will define the degree of efficacy of the information and communication technology. For this, we have to connect versatility with the aims, nature and environment in which the technological development is situated. The efficiency and accessibility are linked fundamentally with the development of both the product and the environment in which it is located.

3.1   Accessibility

Among the internal values that have to be taken into consideration in the evaluation of information and communication technologies (ICTs) are matters related to accessibility. The term "accessibility" has a double meaning, one "physical" and the other "cognitive." The first deals with the convenience of having easy access. From a material perspective, this means that society—as a whole—should be able to use these kinds of tools easily. The second meaning is cognitive. It is linked with the appropriate understanding of the user. This entails that each ICT does not require a high level of training to use it. Therefore, each ICT has to be comprehensible for the large majority of the population.

The Internet is a network that facilitates the exchange, access, and management of information. Its accessibility is linked with the development of the Web. In fact, the Web has increased the accessibility of the Internet in both a physical sense as well as a cognitive sense. Thus, the Internet has made it possible to manage knowledge faster than ever before, in a wider and more complete scope in terms of information, and it is easier to exercise than in previous stages (cf. Floridi 1995, p. 267).[2] For this reason, when the time came to elaborate and construct computers, the accessibility—physical and cognitive—and the versatility (understood as the

---

[2] On Luciano Floridi's philosophy of technology, see Demir (2012).

diversity of the aims) were taken into account. Both the equipment itself and the programs used were orientated towards the economy of means.

Apple was the first company to launch the desktop computer. Soon this move was followed by IBM (cf. Castells 1996, pp. 42–43). Nevertheless, it was the open architecture of IBM computers (i.e., their predisposition to be completely or partially modified) and the possibility of cloning the software by other companies,[3] mainly Asian business-firms, that permitted the large scale spread of desktop use. Nonetheless, the first step was by Apple in 1984, when it began to develop the concept of "easy computing" (cf. Castells 1996, p. 75), making something accessible to a large number of users. The massive spread and use by the population has made "easy use" the main value (besides an attractive external design).[4] For this reason, nowadays Apple brand computers have a growing sector in the market and a more recognizable brand image than their competitors.

A similar evolution occurred in the Internet, at least concerning accessibility. The Internet network was created in 1969. Initially, it had a small group of users in the science and technology fields for two reasons: one physical and the other cognitive. The first reason was its limited physical spread, as the first node, which was found at the University of California in Los Angeles (UCLA) was initially connected to the research institute at Stanford, the University of California in Santa Barbara, and the University of Utah. Then the second initial limitation for accessibility appeared: the control of specialized cognitive contents. This was because a small number of researchers, who were important in their field and possessed a wide knowledge of information science technology, had access to the network. Initially, a lot of training was necessary in order to use this type of information.

With the progressive implementation of the World Wide Web, the technology regarding the Internet was spread to society as a whole. World Wide Web allowed a gradual reduction in the level of knowledge needed to be able to access the Internet. Nowadays, it is only necessary to know a web address (or in some cases how to use search engines). This has promoted the exponential increase in the number of users that demand this technology.

This innovation also attracted the commercial sector to the Internet. That is to say, the network vastly outgrew its roots in the military field and its subsequent boost in the university field. Then the transformation from the "scientific Internet" to the "commercial Internet" occurred. Gradually, the market demand caused a large part of the industrial world to be connected through broadband networks. At that time, this permitted an increase in the number of network services and speed. Without a doubt, it also caused a rise in the number of users.

Flexibility and simplicity pivot around the notion of "accessibility." Commonly, both features happen when someone is developing *hardware* and *software*. It is a characteristic that requires adding the diverse processes that integrate the system, and the efficiency of the whole set depend on it. As Javier Echeverría pointed out, there is a system context: "The operation depends on a number of technical systems

---

[3] As mentioned previously, software connects to the field of the sciences of the artificial.

[4] On aesthetics and technology, see Barker (2012).

strictly interdependent among each other" (Echeverría 1999, p. 95). That is to say, flexibility and simplicity are given in a whole as an organized set.

In order to reach the best levels of accessibility, it is necessary to work on aspects of *hardware* and *software*. Society must access to the devices that permit work and information access in a simple manner; but also programs for information management, and, of course, the Internet. From the perspective of *hardware*, Javier Echeverría points out that "the integration of the different devices that the computer has is very important" (Echeverría 1999, p. 107). Laptop computers, tablets (such as the iPad), and the most developed mobile telephones (i.e., smart phones) allow information access and management in an easier way. We can not only access the internet in enclosed places, like homes and offices, but also in the street, metro, and while taking advantage of the times that we find ourselves in a traffic jam.

Tim Berners-Lee also highlights the need to develop the different mediums for accessing the Internet. In his conception, as opposed to the typical technological notion of "efficiency," accessibility is the most important concept for this technological realm. He considers that it was necessary to launch the mediums to the market, even though they were not great showcases, so everybody could access the Web (cf. Berners-Lee 1999, p. 32).

The other side of the idea of "accessibility" from a physical aspect is linked to the senses. Access to collection, management, and divulgence of information incorporates more senses. Echeverría points out that "'there is a tendency' towards the five senses" (Echeverría 1999, p. 103). At first, the only sense used was sight. A little later sound was also incorporated. The use of touch is already a reality, mostly in devices with portable access such as developed mobile phones (i.e., smart phones) and tablets such as the iPad. Nowadays, researchers are working on incorporating smell and taste.

Also from a physical point of view, the issue of access to the information for disabled people is connected to the accessibility of Internet content by through the different senses (cf. W3C 2013). Tim Berners Lee explains that "all the work on hypertext, graphics, and multimedia languages share concerns about access for all (disabilities)" (Berners-Lee 1999, p. 167).

3.2   Versatility

Within the internal dimension of the analysis of the results, the effects of the action and technological processes are strongly linked to the concept of "versatility." The notion of versatility is connected to the need for technology—as knowledge, human undertaking and product—to accomplish different kinds of functions. Information and Communication Technologies (ICTs) should, in order to be really effective and as efficient as possible, carry out the functions for which they were initially designed.

Certainly there are other options. The first is that the initial objectives were not gained and, therefore, an ICT might serve other functions. Historically, it would not be the first time this occurred. Likewise, a second option is that the function evolved through time, and therefore obtained results not initially expected. A third option is that the users used the same technological product for different functions, although at the beginning it was only intended for one of those functions.

Above all, technology is the key in the infrastructure of the Web. This is because the Internet itself is a means that serves as a technological medium to drive communication. Initially, the functions of the network were related to the profitability of the computing resources as a whole. But with the improvement in the capacity of computer processors, the functions of the Internet were oriented towards purely communicative aims and other goals related to documentation management.

According to a more specific approach, the Internet has the following functions: "1. Remote control of other computers via telnet; 2. File transfers (File Transfer Protocol FTP); 3. Running applications over the Web; 4. Various forms of electronic mail exchange; 5. Publication and consultation of Web pages and other information services available online" (Floridi 1999, p. 67). In this description of concrete actions, one can appreciate the double dimension of the Internet; that feature which is incumbent upon the purely physical aspects—allowing it to operate—, and in terms of content, which make it possible to communicate, inform, and document.

In fact, it is a means of communication that requires a medium. The mediums have the aims of eliminating the distances in the communication, information, and document processes, and facilitating the transmission in a huge way. The innovation, as has been emphasized by Echeverría, is that of the Internet as a vehicle that "tries to integrate the old ways and refer to them" (Echeverría 1999, p. 17). Traditional methods of communication such as written press, radio, and television are integrated in one medium that allows each communication media the use of specific languages.

If you look at it from an epistemological angle, it seems clear that the Internet has introduced important innovations in the organization of knowledge: the reduction of the time-lag between the production and the utilization of knowledge, the promotion of international cooperation and free share of information among researchers and scholars, and the possibility of remote teaching online, etc. (cf. Floridi 1995, p. 267). But this distinct increase in information can also provoke a contrary effect since, as indicated by Luciano Floridi, the Internet increases knowledge, but also ignorance (cf. Floridi 1995, p. 264). This is because what is known as "information contamination" is produced. Indeed, the excess of information can cause a blockage and rejection of the contents, causing the audience to ignore the majority of the messages. This can prevent the objective of increasing the transfer of knowledge from being reached.

Another relevant characteristic of online support is virtuality (cf. Morgan 2010). Together with the physical aspect and the epistemological dimension as content, virtuality is the third aspect. It is located in the orbit of Internet ontology, and in some ways completes the orbit, since it indicates something completely characterized by the network: the virtual component. From an ontological viewpoint, virtuality has its own identity. Through it and within it, we can act. So, the consequences of the actions in the virtual environment are perceived not only in the virtual world, but also in the real world. Indeed, it follows that in the Internet—and in the virtual environment—"you can both communicate and act" (Echeverría 1999, p. 304).

To sum up, the Internet can have a multitude of uses and for this it is both motley and very malleable (cf. DiMaggio et al. 2001, p. 327). Obviously, its multi-use can entail a series of problems, as Manuel Castells points out. In his approach, these are related to the coordination of functions (such functional multi-purpose use can cause attention to be paid to the contradicting functions), the precision of the resources in the aims with the objective of being able to evaluate the effectiveness of the mediums and the execution of the work from a certain level of complexity.[5]

3.3 Efficacy

While dealing with the specific case of information and communication technologies (ICTs), the term "efficacy" is connected to the necessary evaluation of the previously established aims (cf. Simon 1996, p. 6). Efficacy is a fundamental internal value, since it stimulates the achievement of the general aim in technology: the solution to the problems related to reality transformation in a given environment. The Internet is effective *if it works*, i.e., if it achieves its aim. The concept of *efficacy* highlights this teleological focus, because this orientation towards an end is crucial among the internal characteristics of the Internet.

In order to evaluate the level of efficacy in the Internet, it is necessary to recall the aims of the specific design and confirm the extent to which the stated aims have been obtained. In this regard, looking from an internal perspective at the teleological components of the Internet and the Web, one can appreciate this: initially, the technological design has an objective which is generic, because the aim is focused on connecting computers to each other. This connection is obtained through the transmission, processing, and sharing of data.

These computer connections should be made according to the following criteria: (1) the security and resistance of the system;[6] (2) the way for obtaining flexibility; (3) the ability to expand and grow the system; (4) the technological efficacy and management; and (5) the ability to attract users, new and old. These would be able to achieve a large growth in the system.

Commonly, the Web had to solve the problems of format incompatibility for its own design aims. It needs to do so in order to achieve the resource of global information that permitted work in the cooperative way between people and computers. Thus, the connection of thousands of computers on a world scale offers the commercial world a series of unlimited functions. This is possible when there are no difficulties for sharing information (e.g., regarding research). These connections have permitted not only the development once and for all of commerce at a world level, but also the exponential growth in the size of the network, due to the increase in the number of users.

---

[5]Cf. Castells (2001), p 15. On complexity analyzed in dynamic terms see Gonzalez (2013b), pp. 299–311.

[6]On security see Kervalishvili and Michailidis (2012).

Along with the other values, there are the economic values. Economic values have both internal and external characteristics.[7] They deal with three steps: aims, processes, and results. These three economic aspects begin with the design aims of the Internet and the Web. Among these factors are profitability, competitiveness and, especially, yield. At least, these three criteria are incorporated in the aims that serve as a guide to establish a technological design. But enhancing the financial profitability was not a fundamental objective of either the Internet or the Web.

There are also aims that are established from an external environment to the technological field. External demands (social, political, ecological, etc.) emerged that motivated the introduction of variation in the aims and even completely new technological aims. In this regard, one can stress the military aims, which were available at the very beginning of the Internet. These criteria entailed another series of values, such as security, rapid search, and the immediacy of communication. For Tim Berners-Lee, the Web is a fundamental tool of social nature (cf. Berners-Lee 1999). For this key reason, his design included many aspects, such as liberty, responsibility, and security; in addition, it includes ethical as well as political values.

From the point of view of the relationship of means to ends, security is a fundamental element in evaluating the efficacy of the Internet. Security involves the possession of some operative guarantees. This means the inability to access information or documents from those people or computers that have blocked the knowledge of these documents. Therefore, the lack of security can affect personal privacy, and it can also lead to economic crimes. Moreover, it may result in the theft of state secrets, and this could threaten the government of a country. In this sense, three dimensions can be seen in Internet security: (a) the individual level; (b) the institutional sphere; and (c) the governmental dimension.

Within the individual level the most prominent aspects are those related to the privacy and the security of electronic commerce. These values have driven a constant research and development in encryption languages. In this regard, the main goals pursued are: (i) an increase in websites' security and (ii) more guarantees regarding the identification of users of any type of online service.

When the attention goes to the institutional dimension of Internet security, it is necessary to highlight the protocols established by security URLs (Uniform Resource Identifier), which are identifiers that allow access to web resources (or webpages). The security URLs limit access to the content that the users are not authorized for. Besides this, in the institutions listed, the necessity to increase the security in commercial transfers has propelled the development of the https protocol (hypertext transfer protocol secure),[8] which confers the best security for economic transactions and to corporative information systems with sensitive documents.

---

[7] See in this regard Gonzalez (1999). In addition, on the problem of the presence of economic values in science, see Gonzalez (2008). The transition from the value-free Ideal in scientific research to value ladenness in science is considered in Gonzalez (2013a).

[8] Cf. Electronic Frontier Foundation (2010). This Foundation, established in 1990, aims at the defense of civil rights in the electronic media.

Thus, there are a variety of specific aims related to security on the institutional level. These aims have to deal with issues like the security of commercial transfers, the protection of the data that the servers store—which is linked to data protection and privacy at the individual level—, and the attacks of crackers. Usually, crackers try to break the network system of companies and institutions with three possible purposes: (1) the stoppage of their activity; (2) to gain access to certain information; or (3) the challenge of crossing supposedly insurmountable lines.

At a governmental level, the biggest danger is the access and transfer of information from the servers in central and regional governments. States have an enormous quantity of classified documents in their power that are connected to information about civil, economic, political, and military characteristics. The limits of state management of this information are a source of debate at the moment. For this reason, hackers and crackers have made it a challenge to obtain certain types of information, which could cause the disappearance of the political class monopoly on certain types of information. These actions can be evaluated morally from different viewpoints, taking into account several aspects.

Besides security, being a flexible medium is among the most relevant aims of the Internet. Flexibility can refer to both the pure physical dimension (hardware) and the aspects concerning software. In the first case, the method of access to the Internet by means of a basic telephone, which has been valid over the years, has been replaced gradually by other systems. Now there are faster and more stable connections, among them ADSL, Cable Modems, and RDSI. Also other forms of access have appeared through an electric network including via satellite (generally only to download, although there is the possibility of double channels using the protocol DVB-RS).

Along with security and flexibility, another of the aims that the Internet had to achieve is the capacity for growth. From a technological focus, the studies linked to Internet topology examine the size and growth of the network. The growth is reflected in both its physical infrastructure—with the increase in IXPs (Internet Exchange Points)—[9] and the number of users. The increase in users also has influenced their characteristics, so that the profile of Internet users has become wider in scope. In this regard, as previously stated, the first users of the Internet had been connected to the scientific elite in the western universities of the United States. Later, with the development of the World Wide Web, the user profile expanded. In 2013, society's access to information services continues to grow in all geographical areas.

The Internet would continue to be used more, and during the year 2010 more than a fourth part of the people on the planet accessed its services. This growth is

---

[9] The primary role of an *Internet Exchange Point* (IXP) "is to keep local Internet traffic within local infrastructure and to the reduce costs associated with traffic exchange between Internet Service Providers (ISPs). In many developing countries, poor connectivity between ISPs often results in the routing of local traffic over expensive international links simply to reach destinations within the country of origin. IXPs can also improve the quality of Internet services in a country by reducing the delays. Furthermore, IXPs can serve as a convenient hub for hosting value-added and critical infrastructure within a country," Internet Society (2011).

supported by the infrastructures of mobile characteristics (especially in developing countries). North America and Europe are the two parts of the world with the most quotas of penetration, with a rate of above 70 %. Nevertheless, Asian economies have the most users, accounting for more than 40 % of the total users in the world (cf. Fundación Telefónica 2013, p. 35).

After considering the security and endurance of the system, as well as its flexibility and capacity for expansion, the technical efficacy and adequate management of the Internet should be examined. The technical efficacy of the Internet is closely connected with the respect for the coordination of protocols. For this purpose, the presence of an institution that coordinates the technical activity of this medium is necessary. IANA (Internet Assigned Numbers Authority) (cf. IANA 2013) was created by virtue of a contract with the government of the United States for the purpose of managing the central operations that assume the coordination of the Internet on an international level in a way that would work for a global benefit.

In 1998 IANA was replaced by ICANN (the Internet Corporation for Assigned Names and Numbers). ICANN is responsible for assigning the numerical addresses of Internet Protocol (IP), identifiers for protocol and the management or administrative functions of the system of names of generic top-level domains (gTLD), and of country codes (ccTLD), as well as for administrating the root server system. Initially, these services were carried out by Internet Assigned Numbers Authority (IANA) and other organizations under contract with the United States. Nowadays they are the responsibility of ICANN. This association is dedicated to preserving the operational stability of the Internet, promoting the competition, achieving a large representation of world communities on the Internet, and developing the adequate rules for its mission.[10] It is done through processes "from bottom to top" based on consensus (cf. ICANN 2011).

When examining each of the Internet design aims (such as security and endurance of the system, flexibility, capacity for expansion and growth, technical efficacy and adequate management, and its capacity to attract users), one can observe that the achieved effects usually coincide with the goals previously established as the most important within the technological process. What matters for technological success is not merely the achievement of a great diversity of aims (that certainly add complexity to the processes), even though that achievement is really important. What is also needed is the coordination between technological and social goals. In addition, those goals should be related to contents at an international level.

3.4 Efficiency

It is evident that economic values decisively influence the way in which technological processes are developed. This is because the technological design is related to the achievement of the aims with procedures that employ the least number of steps

---

[10]On the role of norms in technology, see Vries, Hansson, and Meijers (2012).

possible, with an economy of means, Therefore, "economic criteria are then influential in the procedure followed in order to make the artifact at stake. Thus, the technological systems are based on an economy of means for achieving the proposed goals. This involves the existence of an *instrumental rationality* in technology" (Gonzalez 1999, p. 74).[11]

If this phenomenon is examined from a methodological point of view, it appears that there is a dynamic component: the evolution of the Internet was also supported in the search for efficiency in the procedure for connecting two or more computers together. That original development required the capacity for interconnections to have increased exponentially. Therefore, its value resides in the capacity for acquisition and transmission of information by someone in an economical (low cost) and rapid manner. There has been an innovation in this field. It was based in optoelectronics and in advanced architectures of commutation and selection of routes. These advances in technology allowed the progressive increase in velocity as well as in the quantity of information transmitted over the Internet; the efficiency of the process was *de facto* enhanced.

This efficiency is observed in another of the pillars that supported the methodological development of the Internet since the changes in the *modus operandi* of the computers (the hardware that allowed access to the network). These changes were conceived from some clear rules with the aim of achieving an "economy of means." In this respect, the innovation in the computers was based on three fundamental design problems: (1) the amount of the information managed; (2) how quickly the information is managed; and (3) the size of the computers. These three priorities are connected to the need to create designs and processes that attend to an "economy of means."

Hence, along with the processes that arise, there are also economic values that influence the knowledge used to make designs (cf. Gonzalez 1998, pp. 103–104). So, within the internal aspect of technology we have economic criteria for technological design such as profitability, competitiveness, productivity, and market quote... Wenceslao J. Gonzalez has pointed out that there are "elements of economic character that have influence on the cognitive component of technology: they are based on criteria of economic rationality and the science of economics itself, and they contribute to specify which technological *objectives* are *preferable*, within those aims that are achievable. This involves the existence in technology of an *evaluative rationality* or rationality of ends, which is under the influence of assessments of economic kind. Thus, the decision regarding the type of artifact to be designed depends on factors of economic character" (Gonzalez 1999, p. 74).

Unquestionably, economic values, such as profitability and competitiveness, have always been present in the design of the Internet and its later evolution. They are elements of intermediation, which appear in the development of the World Wide Web as an alternative to other forms of communication. Those values have a decisive influence on the design of technological goals (at a cognitive level too). On a scale

---

[11] On rationality see also Vogel (2012).

of values, profitability can be considered as the most relevant value to the design of both the Internet and network.

This is because the Internet was an innovation for producing the best benefit in computing since it increases the possibilities for secure and rapid connection. From 1969—the moment ARPANET began to function —[12] to today, one of the main aims of the research and development of the Internet has been to get the best profit from the computing systems. Later in 1996, "Internet2" was created (cf. Internet2 2013). This is a consortium in which the organizations are mainly from the United States, including universities, some big companies, and certain governmental institutions. Its objective is to obtain the most benefit from the Internet through the creation of a telematic network based on the most efficient connections. This involves not only a higher capacity in data storage and faster data transmission but also an increase in the security of this technological infrastructure.

Together with the Internet—an infrastructure directly linked to technological innovation—there is the Web.[13] The Web, which belongs to the software level, was a technological development whose main goal was to obtain greater benefits from the Internet. The network is a computing application that has the goal of solving existing problems in the Internet, and not just the transformation of reality itself. In this sense, the network connects with the sciences of the artificial. Thus, there are background differences between the Internet—understood as a technological infrastructure—and the Web. These differences relate to several levels: logical, epistemological, methodological, ontological, and axiological.[14] Originally, the World Wide Web appeared in the CERN, located in Geneva, which is a technological center for supporting scientific research. There, Tim Berners-Lee noticed the need to increase the benefits of computing systems in a way that it would be possible to have global access to the information. How aim to make computer use more profitable (in order to contribute to improve information access) was his main goal in design. In fact, as Luciano Floridi has noted, "the Internet has made possible a management of knowledge that is faster, wider in scope, more complete in terms of types of information and easier to exercise than ever before" (Floridi 1995, p. 267).

It must be highlighted that financial profit was not a fundamental objective in the design of the Internet. Nor was it in the beginning of the network during its configuration at CERN. Indeed, the Internet was originated in ARPA (Advanced Research Projects Agency), one of the branches of the Department of Defense of the United States of America that searched for a communication system invulnerable to a

---

[12] "The origins of the internet are to be found in ARPANET, a computer network set up by the Advanced Research Projects Agency (ARPA) in September 1969. (...) The first nodes of the network in 1969 were at the University of California, Los Angeles, SRI (Stanford Research Institute), the University of California, Santa Barbara, and the University of Utah", Castells (2001), pp. 10–11.

[13] Technological innovation, in general, and the case of the Web, in particular, can be considered from quite different angles. See, for example, Jordan (2012), Holmquist (2012), Isaacson (2011), and Anderson (2012).

[14] About these aspects of *technology*, see Gonzalez (2005), pp. 8–13.

nuclear attack. So, the technological rationality in the design of the Internet dealt mainly with the value of security in the beginning.

Neither, initially, did the Web have the primary objective of obtaining direct economic benefit. After 20 years of development, the network is founded today in a very different historical context. To a large extent, its purpose is to be useful to humanity. Nevertheless, the Web has resulted in being a computing network—a software—that is financially profitable. In fact, immediately numerous companies noticed that there was business to be had in the World Wide Web.[15] Berners-Lee admitted that the initial target underwent modifications. Soon "it became evident that 'the Web' could splinter into various fractions—some commercial, some academic, some free and some not. This could go against the principal aim of the network: to be a universal hypertext medium accessible for sharing information" (Berners-Lee 1999, p. 76).

The financial profit of the network is not so much due to internal factors (the type of design or the processes of elaboration) as to the growth of the external aspects: its social and public scope. The relevant feature is its competitiveness in comparison to the rest of the communication media. This brings about the introduction of typically commercial elements (advertising of a very diverse nature). According to Berners-Lee, the advantage of the network, as opposed to the rest of the mediums, lies in that, "the Web was not a physical 'thing' that existed in a certain 'place'. It was a 'space' in which information could exist" (Berners-Lee 1999, p. 34). A user of the network can access all the information independently of the place where it is located (with the required equipment as with any other technology) and at the moment its search is conducted.

Although neither the Internet nor the Web were initially designed for home or individual business use (because they were meant for universities, researches, and larger organizations), the technology has had an enormous success regarding its use at home (cf. Berners-Lee 1999, p. 80). Due to its enormously varied and multi-purpose content, its general use has continuously grown in companies, institutions, and homes (cf. Fundación Telefónica 2013; see also Fundación Telefónica 2010). This has led the performance of the Web increasing exponentially, since the value of the Internet is dependent on the quantity of information that can be accessed through this technological instrument as well as the quantity of users that are interconnected to it.

Therefore, from an internal point of view, it is possible to indicate that the Internet and the Web have adapted perfectly. The approach to evaluative rationality has pointed out new aims. These aims, which have been established usually from the environment, have changed over the course of time. The main indicators for

---

[15]While the development of the World Wide Web effectively had a quick support from the business sector; it was not so with the internet. Manuel Castells points out that "the Internet did not originate in the business world. It was too daring a technology, too expensive a project, and too risky an initiative to be assumed by profit-oriented organizations (…) The most blatant illustration of this statement in the fact that in 1972, Larry Roberts, the director of IPTO, sought to privatize ARPANET, once it was up and running (…) After considering the proposal, with the help of experts from Bell Labs, the company refused," Castells (2001), p. 32. On risk see Coeckelbergh (2013).

these conclusions from an analysis of an evaluative rationality are the versatility, accessibility, efficacy, and efficiency (seen here as profitability). Therefore, it is possible to say, from an evaluative rationality point of view that the Internet carries out the main internal values in order to achieve success in this technology.

References

Anderson, P. 2012. *Web 2.0 and beyond: Principles and technologies*. Boca Raton: CRC Press.

Barker, T.S. 2012. *Time and the digital: Connecting technology, aesthetics, and a process philosophy of time*. Lebanon: University Press of New England.

Berners-Lee, T. 1999. *Weaving the web: The past, present and future of the World Wide Web by its inventor*. New York: HarperCollins Publisher.

Castells, M. 1996. *The information age: Economy, society and culture. Volume I: The rise of the network society*. Cambridge: Blackwell.

Castells, M. 2001. *The internet galaxy: Reflections on the internet, business, and society*. New York: Oxford University Press.

Coeckelbergh, M. 2013. *Human being @ risk*. Dordrecht: Springer.

Demir, H. (ed.). 2012. *Luciano Floridi's philosophy of technology*. Dordrecht: Springer.

DiMaggio, P., E. Hargittai, W.R. Neuman, and J.P. Robinson. 2001. Social implications of the internet. *Annual Review of Sociology* 27: 307–336.

Echeverría, J. 1999. *Los señores del aire: Telépolis y el Tercer Entorno*. Barcelona: Destino.

Electronic Frontier Foundation. 2010. http://www.eff.org/pages/how-deploy-https-correctly. Accessed on 29 Mar 2013.

Floridi, L. 1995. Internet: Which future for organized knowledge, Frankenstein or Pygmalion? *International Journal of Human-Computer Studies* 43(2): 261–274.

Floridi, L. 1999. *Philosophy and computing*. London: Routledge.

Fundación Telefónica. 2010. *La Sociedad de la Información en España 2010 siE [10*. Barcelona: Ariel.

Fundación Telefónica. 2013. *La Sociedad de la Información en España 2012 siE [12*. Barcelona: Ariel. Available in: http://info.telefonica.es/sociedaddelainformacion/html/informes_home. shtml. Accessed on 30 Mar 2013.

Gonzalez, W.J. 1998. Racionalidad científica y racionalidad tecnológica: La mediación de la racionalidad económica. *Ágora* 17(2): 95–115.

Gonzalez, W.J. 1999. Valores económicos en la configuración de la Tecnología. *Argumentos de Razón Técnica* 2: 69–96.

Gonzalez, W.J. 2005. The philosophical approach to science, technology and society. In *Science, technology and society: A philosophical perspective*, ed. W.J. Gonzalez, 3–49. A Coruña: Netbiblo.

Gonzalez, W.J. 2008. Economic values in the configuration of science. In *Epistemology and the social*, Poznan studies in the philosophy of the sciences and the humanities, ed. E. Agazzi, J. Echeverría, and A. Gómez, 85–112. Amsterdam: Rodopi.

Gonzalez, W.J. 2013a. Value ladenness and the value-free ideal in scientific research. In *Handbook of the philosophical foundations of business ethics*, ed. Ch. Lütge, 1503–1521. Dordrecht: Springer.

Gonzalez, W.J. 2013b. The sciences of design as sciences of complexity: The dynamic trait. In *New challenges to philosophy of science*, ed. H. Andersen, D. Dieks, W.J. Gonzalez, Th. Uebel, and G. Wheeler, 299–311. Dordrecht: Springer.

Holmquist, L.E. 2012. *Grounded innovation: Strategies for creating digital products*. San Francisco: Elsevier.

IANA. 2013. http://www.iana.org/about/presentations/davies-atlarge-iana101-080929.pdf. Accessed on 29 Mar 2013.

ICANN. 2011. http://www.icann.org/tr/spanish.html. Accessed on 17 Oct 2011.

Internet2. 2013. http://www.internet2.edu/. Accessed on 30 Mar 2013.

Internet Society. 2011. http://www.isoc.org/internet/issues/ixp.shtml. Accessed on 13 Oct 2011.

Isaacson, W. 2011. *Steve Jobs*. New York: Simon and Schuster.

Jordan, J.M. 2012. *Information, technology and innovation. Resources for growth in a connected world*. Hoboken: Wiley.

Kervalishvili, P.J., and S.A. Michailidis (eds.). 2012. *Philosophy and synergy of information: Sustainability and security*. Georgia: Ios Press.

Morgan, M.S. 2010. 'Voice' and the facts and observations of experience. In *New methodological perspectives on observation and experimentation in science*, ed. W.J. Gonzalez, 51–70. A Coruña: Netbiblo.

Oosterlaken, I., and J. Hoven (eds.). 2012. *The capability approach, technology and design*. Dordrecht: Springer.

Simon, H.A. 1996. *The Sciences of the artificial*, 3rd ed. Cambridge, MA: The MIT Press.

Vogel, M. 2012. *Media of reason: A theory of rationality*. New York: Columbia University Press.

Vries, M.J., S.O. Hansson, and A.W.M. Meijers (eds.). 2012. *Norms in technology*. Dordrecht: Springer.

W3C. 2013. http://www.w3.org/WAI/. Accessed on 29 Mar 2013.

Part II
Rationality and Responsibility in Technology

Chapter 4
Rationality in Technology and in Ethics

Carl Mitcham

Rationality is a generally acknowledged good, yet one that takes different forms in different contexts–and is sometimes qualified. The primary effort here is to reflect on some tensions between rationality as manifested in technology and in ethics. Initially, however, it is appropriate to venture some general observations about rationality.

4.1 Rationality and Its Diverse Contents

That rationality (which must be distinguished from the Enlightenment philosophy of rationalism) takes a variety of forms can easily be indicated. According to its English etymology and the lexical description from the *Oxford English Dictionary*, the term derives from the post-classical Latin *rationalitas*, faculty of reasoning. It is an abstract substantive from the adjective "rational," indicating the exercise or possession of reason; a closely related abstraction is reasonableness. Its deeper root is the Latin *ratio* (often used in Roman philosophy to render the Greek *logos*), with meanings that range, according to the *Oxford Latin Dictionary*, from the act of reckoning or calculating, especially financial accounting, and proportion or relation, to the act or process of reasoning or working out, an explanation or reason, a descriptive account, the exercise of reason, an affair or business, a plan of action, guiding principle or rule, and method or means.

It is important to note, however, that in non-European languages etymology and denotations can be quite different. In classical Chinese philosophy, for instance, there is no word that can be translated as "rationality"; in modern Chinese the term

C. Mitcham (✉)
Liberal Arts and International Studies, Colorado School of Mines,
Stratton 301, Golden, CO 80401, USA
e-mail: cmitcham@mines.edu

© Springer International Publishing Switzerland 2015   63
W.J. Gonzalez (ed.), *New Perspectives on Technology, Values,
and Ethics*, Boston Studies in the Philosophy and History of Science 315,
DOI 10.1007/978-3-319-21870-0_4

most commonly translated as "rationality" is *lixing*, which consists of two compound characters. The first, *li*, is composed of the radical for king or ruler and the character for inside, village, or neighborhood; the second, *xing*, is constructed from the radical for heart and the character for birth or life.

Few twentieth century English-language encyclopedias of philosophy have entries on rationality itself, but often discuss rationality in some context. Neither the influential *Encyclopedia of Philosophy* (1967) nor the contemporary online *Stanford Encyclopedia of Philosophy* carry entries on rationality; the *Routledge Encyclopedia of Philosophy* (1998) only discusses the topic in distinct entries on "Rationality and Cultural Relativism," "Rationality of Belief," and "Rationality, Practical." The *Encyclopedia of Philosophy Supplement* (1996) added a short entry by Paul Moser on "Rationality" (Moser 1996; reprinted unchanged in the *Encyclopedia of Philosophy*, second edition, 2005) that likewise distinguishes epistemic and practical rationality. Then with regard to practical rationality Moser further distinguishes instrumental versus substantive rationality; the former concerns the selection of effective means to achieve some predetermined end, the later with the identification of proper ends. While acknowledging that the modern marginalization of substantive notions of rationality (witness the focus on instrumental notions in decision theory and related research programs), Moser elaborates on four conceptions of egoistic, perfectionist, utilitarian, and intuitionist rationality distinguished by William Frankena (1983), adding a fifth, relevant-information conception. Moser argues that rationality and morality may or may not conflict, depending on the precise interpretation of each of these five types of ends. As if confirming the primacy of instrumental rationality, the *Oxford Handbook of Rationality* (2004) discusses substantive rationality at length in only one of 22 chapters.

Related to the instrumental/substantive distinction is another between rationality and reasonableness. According to a brief account by Alan Gewirth in the *Encyclopedia of Ethics* (1992 and 2001), persons are rational if they choose the most efficient means to their ends, whatever they may be; reasonable if they maintain a certain equitable relationship between themselves and others, that is, consider ends from an impartial perspective. "Thus, reasonableness is directly a moral quality, while rationality is often nonmoral, and may even be immoral if the agent's ends are exclusively self-interested" (Gewirth 1992, p. 1069). Although the specific terminological distinction is not widely adopted, the distinction itself is real. Purely instrumental rationality can be at odds with moral reasoning.

A parallel trajectory of attention can be found in the social sciences. There are no entries on rationality per se in either the *Encyclopedia of the Social Sciences* (1935) or the *International Encyclopedia of the Social Sciences* (1968). Rationality is finally granted thematic treatment in the *International Encyclopedia of the Social Sciences*, second edition (2008) with an entry by philosopher Paul Weirich. After a brief general introduction, three-quarters of the main body of this short (seven column) entry are devoted to rationality as some version of instrumental utility maximization. Beyond Weirich, it is possible to distinguish at least four distinctive research programs dealing with rationality. One is the rational choice and decision theory research of economists and others who focus on instrumental rationality and seek to determine its procedural norms and applications. In the selection of means, what legitimately counts as evidence and how; how is the means-ends relationship most effectively internalized in decision making. Two is research by psychologists and others who seek to identify how people express and reveal preferences (ends) as well as actually make choices and decisions, never mind what is the most effective way to make such choices and decisions. Three is research by psychologists and anthropologists on the evolutionary origins of rationality. What are the biological or genetic foundations of reasoning and how has it evolved in conjunction with other features of human development. Fourth is research that seeks to evaluate the degree to which real-world decision making accords with decision theoretical norms along with what might be done to meliorate failures to do so. Leading insights of philosophical importance that cross these diverse research programs on instrumental rationality can be found in, e.g., the work of Herbert Simon (1969 and 1983) and Daniel Kahneman (2011).

In the present context, then, what may be emphasized from the start – on the basis of both philosophical and social science discourse – is the potential for rationality to be in tension with if not opposed to other aspects of human experience. Insofar as rationality denotes a dependence on reason it can be contrasted with or opposed to revelation, necessity, intuition, perception, beauty, emotion, and more. For present purposes the most salient tension is one between rationality and morality or ethics, the two most prominent normative dimensions of human experience. (Qualification: The terms "morality" and "ethics" are sometimes treated as interchangeable, although in technical parlance "morality" refers to behavior and "ethics" to critical reflection on behavior.)

In ordinary language it is not uncommon to hear the rationality/ethics tension expressed in one of the following templates: "X is rational but not ethical" or "Y is ethical but not rational." Instances of X that fit the first case include the atomic bombing of Hiroshima and Nagasaki, the torturing of terrorists, transferring profits to tax haven jurisdictions, and more. Instances of Y in the second case include turning the other cheek, outlawing the death penalty, not cheating on one's taxes when one can get away with it, and more. Of course, in each of these cases arguments can be made to harmonize rationality and morality, but the point is that such arguments need to be made. *Prima facie*, it makes more sense to say there is an opposition between what is reasonable and what is moral.

In ordinary language, however, it is difficult to think of instances in which rationality is in such obvious tension with or opposed to technology. That is, given the linguistic templates, "X is rational but not technological" or "Y is technological but not rational," it is more difficult to imagine substitutes for X or Y. In the first case perhaps the best candidates are actions that are often characterized as performances, such as promising or loving; but to describe these as rational also sounds a bit odd. In the second case Y could be an inefficient or bad technology. Example: "That Rube Goldberg machine is technological but not rational."

At the same time there is a deep sense in which any notion of rationality implicates some notion of the good. Rationality is not self-justifying. Although one can say it is irrational not to be rational, this is simply a tautology. More substantive is an argument that it is immoral or unethical not to be rational. Notice too how it is not equally the case that any commitment to ethics is rational; ethical commitments have been based on appeals to revelation, tradition, experience, and power, although some proponents defend such appeals as exhibiting their own distinctive forms of rationality. Even the Kantian effort to derive morality from rationality requires some prior sense of rationality as good.

One classical way of conceiving rationality as good is to understand it as the perfection or virtue of a faculty of the mind sometimes called the intellect. In Plato and Aristotle the functional perfection of the mind or intellect (*nous*) is intelligence (*noesis*), which could also be translated as rationality. Insofar as rationality is a perfection of the intellect, this implies a strong relationship between rationality and intelligence. Even today we often say that an intelligent person is rational or that a rational person is intelligent; so-called intelligence tests commonly measure skills of reasoning. In the analogy of the divided line, however, Plato distinguishes between *noesis* (intuitive reasoning) and *dianonia* (discursive reasoning) that foreshadows one between substantive and instrumental rationality thus suggesting two types of intelligence. Additionally, contemporary philosophically relevant work by Simon, Kahneman, and others has established that otherwise apparently quite intelligent people can often make irrational choices, thus implicating a need to distinguish intelligence and rationality.

From this brief review of rationality and its multiple manifestations, the points most germane to further reflection on rationality as manifested in technology and in ethics may be summarized as follows: We desire to be rational only insofar as we see rationality as a good, instrumentally or substantively. The basic argument for the pursuit of technological rationality is thus an ethical one. At the same time, arguments exist for making ethics itself more technological. Both arguments deserve explication and examination.

The complexity of the arguments at issue dictate that the present reflection be no more than a preliminary foray. Nevertheless, the ultimate aim is to defend the thesis that ethical rationality trumps technological rationality, both in practice and in theory–and for good reasons, indeed for the good.

The complexity here derives in part from the fact that beyond the contextual differences between technology and ethics, technological rationality and ethical rationality themselves exhibit multiple context-related forms. As a result, what follows will consist primarily of two case studies.

The first case highlights technology by considering how engineering has attempted to incorporate ethics into professional self-understandings. Engineering is thus taken as a central aspect of technology. As has been argued on other occasions, technology is constituted by the systematic making and using of artifacts, including all the artifacts themselves; engineering is the design and construction of artifacts (Mitcham 1994; Mitcham and Schatzberg 2009; see also McCarthy 2009, and Blockley 2012). Engineering ethics thus constitutes a specific effort to build a bridge between technological and ethical rationality, starting from the side of technology.

The second case considers some relations between ethics and policy. The relationship of policy to technology will be developed in due course. But insofar as policy can also be understood a kind of technologization of politics, the ethics-policy relationship is another instance of bridge building between two rationalities, this time beginning more from the side of ethics.

The two case studies are followed by some general reflections that relate the two cases in the form of comments on technology and democratic society.

4.2 Engineering and Ethical Rationality

The first case study considers the role of ethics in engineering. Because it is difficult to think engineering in general, the present historico-philosophical reflection further contextualizes engineering with a focus on how it has developed and been practiced in the United States.

The classic definition of modern or scientific engineering as practiced in North America derives from one formulated in conjunction with establishment of the British Institution of Civil Engineers (ICE). In 1818, at the first ICE meeting, H. R. Palmer, while lamenting the absence of any organization to serve as a "source of information or instruction for persons following or intending to follow the important profession of a Civil Engineer," described the engineer as "a mediator between the Philosopher and the working Mechanic," that is, one who learns the principles of nature from natural philosophy "and adapts them to [human] circumstances" while the "working mechanic … brings [the engineer's] ideas into reality" (Institution of Civil Engineers, January 2, 1818). Ten years later, in conjunction with the application for a royal charter, ICE president Thomas Tredgold more carefully defined engineering as the "art of directing the great sources of power in nature for the use and convenience of [human beings]" (Tredgold 1828).

This definition significantly leaves out some issues while including others. Tredgold erases any explicit reference to science, although it had been present in Palmer's description. But then Tredgold underlines Palmer's notion of adaptation to human circumstances by referencing human "use and convenience" as the ethical end. "Use and convenience" is a semi-technical term associated with the development of utilitarian philosophy during the same period. In his *Enquiry Concerning the Principles of Morals* (1751), David Hume observed that all art or well-designed making is oriented toward human "use and convenience." Notions of use and convenience subsequently came to play important roles in classical economics, from Adam Smith on.

However, even within the framework of modern human commitments to this world and material progress, use and convenience are subject to divergent interpretations. The social context in which this-worldly ends are to be pursued remains open and debatable. Although Tredgold and the Institution of Engineers viewed use and convenience as a non-problematic purpose for engineering, it is remarkable that no subsequent engineering ethics code affirms this end. Use and convenience have never functioned in quite the same way in engineering as health in medicine or justice in law; use and convenience tend to function more as what might be termed "ex-forming" than as informing ideals. To suggest the same point in different words, they operate at such a high level of generality as to require interpretation and application.

Ethics will necessarily come into play in any related discussion of the particular meaning of use and convenience or the contexts in which use and convenience are to be pursued. Useful to whom (factory owners, investors, clients, consumers)? Convenient for whom (workers, sellers, purchasers)?

In relation to the interpretation of use and convenience, it is important to draw attention to the relatively recent development of engineering as a profession. Human beings have since antiquity undertaken projects that may, from a modern vantage point, be interpreted as engineering – witness, for example, Egyptian pyramids, Roman aqueducts, the Brihadishwara Temple in India – just as science has projected its history back to the Greeks and beyond. Nonetheless, in the West the first engineers as such did not appear until the Renaissance. It was at this historical juncture that a systematic or scientific approach to questions of what works and why in both structures and machines began to displace the earlier trial-and-error thinking of artisans and architects. Indeed, Galileo Galilei's *Two New Sciences* (1638), which adopts a scientific approach to practical problems and structural analysis, is widely regarded as a landmark text in the history of engineering.

If the birth of modern science as an institution can be dated from the founding of the Royal Society in 1660, engineering as a profession is best dated from a century later with formation of the Society of Civil Engineers in 1771. Since its distinctly modern emergence, there have developed three theoretical ideals in engineering ethics–which in effect constitute three interpretations of use and convenience.

4.2.1 Use and Convenience Through Obedience to Authority and Company Loyalty

The first theory of engineering ethics grants to the market the determination of use and convenience and thus makes engineers subordinate to corporations that employ them. Their fundamental obligation is to obey or to be loyal to organizations or firms in which they work.

Engineering as a profession initially took distinctive form in the military. An "engineer" was originally a soldier who designed military fortifications and/or operated engines of war such as catapults. The first engineering schools were founded by governments and closely linked with the military; one early example was the Academy of Military Engineering at Moscow created by Czar Peter the Great in 1698. What is often taken as the archetypical engineering school is the École Polytechnique, founded at Paris in 1794, which became a military institution under Napoleon Bonaparte. In the United States the first school to offer engineering degrees was the Military Academy at West Point, founded in 1802. Within such contexts, the over-arching duty of engineers, as with soldiers of all types, was to obey orders.

During the same period as the founding of professional engineering schools a few designers of "public works" began to call themselves "civil engineers"–a term that continues in some languages to denote all non-military engineers. The creation of this civilian counterpart to engineering in the armed forces initially gave little reason to alter the fundamental engineering obligation. Civil engineering was simply peacetime military engineering, with use and convenience replacing protect and destroy, and engineers remained duty-bound to obey those for whom they worked, whether some branch of the government or a private corporation.

The late eighteenth and early nineteenth centuries also witnessed formation of the first professional engineering societies as organizations. But none of the original associations included any formal code of ethics. Formal ethics statements had to wait until the early twentieth century. On analogy with physicians and lawyers, whose codes prescribe a fundamental obligation to patients and clients, the early codes of conduct in professional engineering–such as those formulated in 1912 by the American Institute of Electrical Engineers (later to become the Institute of Electrical and Electronic Engineers or IEEE) and in 1914 by the American Society of Civil Engineers (ASCE)–defined the primary duty of the engineer to serve as a "faithful agent or trustee" of an employing company.

The implicit argument behind this view of engineering ethics is that engineering best produces the general good of use and convenience through corporate subservience and the free market. But insofar as economic competition and consumer choice can produce mistreatment of workers, poor quality products for consumers, and environmental degradation, this ethical justification of engineering may be questioned as producing something other than simple use and convenience.

4.2.2 Use and Convenience Through Technocratic Leadership and Efficiency

At odds with both the implicit code of obedience and the explicit code of company loyalty is the ideology of leadership in technological progress through pursuit of the ideals of technical perfection and efficiency. During the first third of the twentieth century in the United States this vision of engineering activity spawned the technocracy movement or a belief that engineers should be given political and economic power. Although never explicitly articulated in the form of a code of conduct, it has influenced how engineers and the public think about the profession. Economist Thorstein Veblen, for example, argued that if engineers were freed from subservience to business interests their own standards of good and bad, right and wrong, would lead to the creation of a more sound economy and better consumer products (Veblen 1921).

The technocracy ideal was formulated during the same historical period that governments in Europe and North America were establishing independent agencies to regulate transport, construction, communications, foods, pharmaceuticals, banking, and more. In all such cases technical experts from science and engineering were given governing responsibilities to oversee the operations of public and private activities, with an aim of increasing perfection, efficiency, and safety. The pursuit of efficiency also spilled over from industrial to social life. Engineering efficiency became a model for enhancing personal use and convenience in managing one's health, finances, education, and general comportment (Alexander 2008).

There are good reasons to practice some degree of technocratic leadership and efficiency. Certainly the subordination of production to short-term money making with little concern for product quality is not desirable in the long run, and inefficient or wasteful processes can readily be described as wrong. Moreover, in a highly complex technical world it is often difficult for average citizens to know what would be in their own best interests. Efficiency is not always adequately promoted by consumer pull in imperfect markets; it sometimes requires push from technical experts.

Nevertheless, when technical decision making becomes a formalized process, it is easily decoupled from general human welfare. Not only can regulatory agencies be captured by the industries or activities they are supposed to regulate (a version of the principal-agent problem) but the pursuit of efficiency is not always compatible with personal happiness. Concepts of technical perfection and efficiency virtually require the assumption of clearly defined boundary conditions that per force can exclude important and relevant factors, including legitimate psychological, environmental, and human concerns.

Technical efficiency entails minimizing inputs to achieve desired outputs or maximizing outputs from given inputs – or both. It therefore hinges completely on how inputs and outputs are framed, which is not a strictly technical matter. In recognition of such objections there has developed a third theory of engineering ethics, that of social responsibility.

4.2.3 Use and Convenience Through Public Safety, Health, and Welfare

The World War II mobilization of science and engineering for national purpose and the North American post-war economic recovery caused a provisional suspension of the tension between technical and economic ends, efficiency and profit, that had come to light in discourse associated with technocracy. But the anti-nuclear weapons movement of the 1950s and 1960s, in conjunction with the consumer and environmental movements of the 1960s and 1970s, brought tensions again to the fore and provoked some engineers to challenge national and corporate or business direction as well as the technocratic ideal. In conjunction with a renewed concern for democratic values–especially as a result of the civil rights movement–this led to new ideals for engineering.

The emergence of a new theory of social responsibility has a complex history that draws on multiple dissatisfactions with both the first and second theories. In the United States the seeds of transformation were planted immediately after World War II when in 1947 the Engineers' Council for Professional Development (ECPD which later became the Accreditation Board for Engineering and Technology or ABET) drew up the first trans-disciplinary engineering ethics code, committing the engineer "to interest himself [or herself] in public welfare." Revisions in 1963 and 1974 strengthened this commitment to the point where the first of four "fundamental principles" required engineers to use "their knowledge and skill for the enhancement of human welfare," and the first of seven "fundamental canons" stated that "Engineers shall hold paramount the safety, health and welfare of the public ..."

This third theory addresses many problems with the other two, and has been widely adopted by the professional engineering community, in the United States and elsewhere. It also allows for the retention of desirable elements from prior theories. For instance, obedience or loyalty remain, but within a larger or more encompassing framework. Now the primary loyalty is not to some individual or corporation but to the public as a whole. Leadership in technical perfection and efficiency likewise remains, but is explicitly subordinated to the public welfare, especially in regard to health and safety.

There are nevertheless questions to be raised with regard to the theory that use and convenience are to be achieved through engineering responsibility for public safety, health, and welfare. One concerns whether engineers qua engineers really have any privileged knowledge with regard to safety, health, and welfare. This issue will be returned to below.

4.3 Ethics, Policy, and Rationality

A second case study considers some relations between ethics and policy. Adapting a distinction prominent in discussions of science policy, it is possible to distinguish two relationships. One focuses on ethics for policy, another on policy for ethics. The former concerns how to bring ethics to bear in arenas of policy formation and decision making, the latter on what policies might best be used to promote or develop ethics.

4.3.1 Policy Itself

Before taking up these two relationships consider the concept of *policy* itself, increasingly understood (in the U.S. context) as scientifically and technologically based guidance for behavior that achieves rational outcomes. As such, policy constitutes a kind of technological rationality closely related to engineering; it might even be described as decision engineering.

The under-discussed demarcation problem in policy studies concerns the distinction between policies in this sense from other aspects of human affairs, such as politics, law, rules, plans, designs, principles, or ethics. One scholar prominent in developing the notion of policy as counterpoint to politics was the interdisciplinary American social scientist Harold D. Lasswell. In his classic paper, "The Policy Orientation," Lasswell wrote:

"The word 'policy' is commonly used to designate the most important choices made either in organized or in private life. We speak of 'government policy,' 'business policy,' or 'my own policy' regarding investments and other matters. Hence 'policy' is free of many of the undesirable connotations clustered around the word political, which is often believed to imply 'partisanship' or 'corruption'" (Lasswell 1951, p. 5).

Lasswell and others developed a method to formulate and assess policy by means of what he termed the "policy sciences" to support a systematic, interactive sequence that includes goal clarification; detailed empirical assessment of the situation in which a goal is to be pursued; the careful weighing of alternative courses of action; and the continuous evaluation and selection of optimal means for carrying out a selected course of action. Any science becomes a policy science insofar as it contributes to the policy making process.

Policies in this sense are supposed to be based not in politics (with its characteristic appeals to tradition, power, or majority rule) but in science (with its appeals to empirical evidence or theoretical adequacy); when not legally codified, they lack the force of law and the constitutive character of rules while still being able to guide law enactment and rule formulation; they thus serve as plans or designs both for subsequent decision making as well as for particular actions or artifacts; finally, they function at a level of abstraction intermediate between specific decisions and general principles. They can either include ethics or themselves constitute a kind of (primarily consequentialist) ethics.

Genesis of the term "policy" in this distinctive sense can be traced to the late 1800s and proposals for supplementing legislation on the basis of tradition, interest-group power, and common sense politics with the creation of laws and regulations focused on meeting a public interest while appealing to scientific evidence and analysis. For example, neither tradition, nor power politics, nor ethics, nor common knowledge were able in the eighteenth century to determine how best to supply safe drinking water to expanding urban centers. Instead, an ethical commitment to public health was increasingly dependent on knowledge supplied by scientists such as Antonie van Leeuwenhoek (1632–1723), who invented the microscope and discovered microbial life in ostensibly pure water, along with the technical skill of engineers such as John Gibb (1776–1850)–a founding member of the Institution of Civil Engineers – who designed the first large-scale water filtration system in Paisley, Scotland. Discoveries such as those of epidemiologist John Snow (1813–1858) regarding cholera transmission via water contamination and formulation of the germ theory of disease by biologist Louis Pasteur (1822–1895), further enhanced the ability of science and engineering to trump traditional politics and ethics alone in governmental efforts to protect the public from unsafe water, transport, structures, and food.

The typical process is for a legislative body to pass a law mandating the creation of, e.g., safe drinking water standards, then delegate determination of the standards themselves to the use of science and technology by some specified agency. Furthermore, to support evidence-based policy making, many modern governments have found it necessary to fund scientific and technological research through state research institutions, grant support to independent scientific institutions or researchers, and/or tax incentives that encourage private enterprises to undertake relevant research. This is obviously another version of second-stage engineering ethics ideals of technocracy—or, perhaps more accurately, what should now be termed "scientotechnocracy" or "technoscientocracy."

Involvement by scientific and technical experts in effective policy making nevertheless creates the previously mentioned principal-agent problem: How can the principal (in this case, the state) be sure that expert agents (scientists and engineers) share the principal's goals? It is precisely such concerns, especially in democratic contexts, that have promoted discussions that may be classified under the rubric "ethics for policy."

4.3.2 Ethics for Policy

Ethics for policy is focused on how to bring ethical practices, principles, reasoning, and considerations to bear in areas of public policy deliberation, design and implementation. In practice the focus is largely on morals, or the inculcating of behavioral norms acceptable to decision makers and their constituents, and rarely on ethics, that is, critical reflection on such norms. Ethics for policy typically argues, for instance, in support of loyalty to established authorities, protection of confidential information acquired during the performance of professional duties, and avoiding conflicts of interest. Again, comparisons with first-stage engineering ethics codes should be obvious.

Consider again the case of drinking water. While even preliterate peoples understood the importance of an ample quantity of water to support human habitation, water quality was traditionally assessed simply on the basis of taste (non-salinity) and aesthetics features such as visual appearance (absence of turbidity) or smell. Microbiology made possible scientific and technological efforts to avoid water-born diseases. In the United States, during the late nineteenth and early twentieth centuries, safe drinking water standards were progressively recognized as important public health issues. In 1912 the newly established U.S. Public Health Service (PHS, as the reorganization of related agencies that can be traced back to 1798) was mandated to "study and investigate the diseases of men and conditions influencing the propagation and spread thereof, including sanitation and sewage and the pollution … of navigable streams and lakes" (Dupree 1957, p. 270). Two years later the PHS issued the first standards for the bacteriological quality of drinking water, which were revised and expanded in 1925, 1946, and 1962.

By the late 1960s, however, scientific research on drinking water was implicating not just infectious disease-causing pathogens but also industrial chemicals as causal factors in non-infectious birth defects, child developmental disabilities, and illnesses such as cancer. These technoscientific chemicals were increasingly finding their ways into water sources. To examine and regulate the associated risks, the U.S. federal government in 1970 established the Environmental Protection Agency (EPA) and followed up with a series of legislative acts explicitly addressing the need for further research and rulemaking: the Clean Water Act (1972) and the Safe Drinking Water Act (1974) with amendments (1986 and 1996). An official EPA twenty-five year history of the results concluded, "To continue learning about the health effects of known and/or regulated contaminants, and to begin studying emerging contaminants (e.g., newly discovered microbes, perchlorate), it will be imperative that the public and private sectors work together to more effectively and efficiently conduct sound scientific research in the future" (Environmental Protection Agency 1999, p. 35).

When, on the basis of its rulemaking authority and scientific evidence, agencies such as the EPA mandate compliance with new technical standards, the principal-agent problematic emerges. Since the rulemaking is dependent on scientific or technological knowledge that is seldom obvious to the general public, and in some instances the evidence itself may be unclear or conflicting, how can the government or public be sure that its technoscience and their technoscientific experts are not compromised by conflicts of interest or their own personal commitments?

One effort to respond to this question has been to create codes of ethics for technoscientific experts in government service who are involved in policy assessment, implementation, and other forms of public policy making. For instance, in 1958, in an effort in part and in effect to promote ethics for policy analysis and rulemaking, the U.S. Congress set forth a "Code of Ethics for Government Service," stipulating that any civil servant should, among other things, "Make no private promises of any kind binding upon the duties of office, since a Government employee has no private word which can be binding on public duty."

Such efforts to create codes of conduct to guide scientists and engineers in relation to their engagements with public policy are clearly related to the ethics codes of professional engineering societies – but with a difference. With engineering ethics codes, the impetus and development was internal to the profession, even if the effect was to adopt external ideals. By contrast, here the efforts themselves are external to the profession and so constitute what may be described as a "policy for ethics." Especially is this the case insofar as such codes of ethics are associated with some systematic study of how best to promote behavior that achieves outcomes, and is thus instrumentally rational.

4.3.3 Policy for Ethics

Like science for policy, ethics for policy is focused on rationally influencing and shaping public decision making. Like policy for science, policy for ethics explores the need for and the most rational approaches to providing ethical guidance. That is, policy for ethics examines the nature, source, legitimacy, and promotion–including funding support–of such ethics for policy. In the broad sense, given the general role of ethics in society, policy for ethics includes questions concerning the best ways to promote or develop moral behavior and other dimensions of ethics, not just among policy professionals and public servants but among all citizens.

In one sense this issue is as old as Plato's *Republic* and Aristotle's *Politics*, both of which considered how best to structure the *polis* so as to cultivate virtue among citizens, although neither philosopher considered civic virtue as subject to modern scientific and technological policy determination. In modern secular societies, increasingly informed by and dependent on technoscience, many traditional institutions for instilling and enforcing norms–such as families, local communities, religious and educational institutions–have been significantly weakened, just at the point at which extended technological powers demand greater conscious reflection than ever before. In consequence, appeals are made to social scientific and other forms of expertise to help determine how to cultivate morality.

From the middle of the twentieth century, as the physical conditions of the human lifeworld were transmogrified through technoscience, a number of assumptions regarding the norms of human conduct underwent inversions. For example, when many children failed to live to adulthood and the human population was mostly stable over long periods of time, unlimited procreation was an obligation reinforced by natural human inclinations. Once advances in public health significantly increased the survival rate of children and lengthened the average human life span, procreation became an action subject to reflective delimitation. Indeed, through the technologies of birth control, what had once been more a behavior than an action, has increasingly come to call for conscious decision making and the taking into account of more factors than had ever previously been the case–entailing what has been termed a duty *plus respicere* (Mitcham 2011).

Another example: As long as farming was done on small scales with technics inherited from tradition, it made sense to cultivate all available land, as had been done for generations–although small scale alone does not insure against major failures to appreciate the limits of nature (Diamond 2005). As new technologies and scales were introduced, the conscious mediation of the agricultural extension agent became an almost necessary adjunct, and in some cases even the de-cultivation of arable land became a newly appropriate norm.

Insofar as humans experienced few lifeway options, they did not have to worry about which choices might be best; as options proliferated, they increasingly were encouraged to consider economic and psychological factors when making decisions. The pattern of increased need for research and reflection manifested in the public sphere design, construction, and operation of urban water systems was imported into many spheres of personal decision making.

The situation of medical care provides still another vivid illustration. As medicine has become increasingly empowered by life science engineering, public policy has been stimulated to develop formal ethical engagement mechanisms for decision making regarding utilization of the related therapeutic technologies. The establishment of ethics advisory committees–known in the United States as Institutional

Review Boards (IRBs) and elsewhere as Ethical Review Boards or Independent Ethics Committees–has been a widely adopted initiative made on the basis of appeals to social science assessments of public need and to ethical expertise instead of to interest group politics; as such these too may be conceived as a form of policy for ethics. The U.S. National Research Act of 1974 defined the structure of IRBs and requires them for all research directly or indirectly funded by the Department of Health and Human Services with the aim of introducing an ostensibly non-partisan analysis of options along with enhanced and broadened ethical reflection.

During the last third of the twentieth century a host of similar initiatives emerged that, insofar as they have purported to be science- or reason-based programs for action to guide decision making toward the effective development and utilization of ethical expertise in human affairs, could be classified as part of a policy for ethics spectrum. Among such initiatives are the following:

1. Beginning in 1974 in the United States (and subsequently in many other countries), creation of a series of national bioethics commissions and committees to bring ethical expertise to bear at a public level in discussions of challenges emerging from advances in biomedicine (Briggle and Mitcham 2005).

2. Establishment in 1975 of the Ethics and Values in Science and Technology (EVIST) program at the U.S. National Science Foundation (with a related program at the National Endowment for the Humanities), which in the 1980s morphed into the Ethics and Values Studies program, to fund "studies of ethical and value aspects of the interactions between science, technology and society" as a new research area (Hollander and Steneck 1990).

3. Shortly after its formal inception in 1986, the Human Genome Project started to include Ethical, Legal, and Social Implications (ELSI) research into government-funded efforts to map and sequence the human genome (Langfelder and Juengst 1993). (In Europe ELSI research is more commonly termed ELSA or Ethical, Legal, and Social Aspects research.)

4. Following exposure of a series of medical research misconduct cases, in 1989 the U.S. National Institutes of Health began to require that all graduate students on training grants receive education in responsible conduct of research (*NIH Guide* 1989).

5. In 2003 the U.S. National Nanotechnology Initiative (NNI) includes a "post-ELSI" program to promote public engagement and the integration of social sciences into nanoscience and engineering, which led to establishment of two Centers for Nanotechnology in Society (one at Arizona State University and another at the University of California Santa Barbara).

6. In January 2010 the National Science Foundation, under mandate from 2007 legislation by the U.S. Congress, began to require that any student or postdoc who receives NSF support have training in responsible conduct of research.

Both ELSI and post-ELSI activities are significant developments of policy for ethics, in terms of scale and the inclusion of the social sciences and even the humanities in the policy process. In notable contrast to the ELSI program, which has been criticized precisely for its failure to inform policy as specified in its mandate (Fisher 2005), post-ELSI programs call for integration of the social sciences into decision making in order to influence the direction of research, development, and commercialization (Fisher and Mahajan 2006).

4.3.4   Policy Limits

Ethics for policy and policy for ethics are complementary efforts to make present something that is absent. In this the notion of ethics policy is continuous with the arguments of great nineteenth century radical philosophers such as Karl Marx (1818–1883) and Friedrich Nietzsche (1844–1900). But whereas Marx sought to reinsert ethics into the lifeworld through political activity and Nietzsche through a great-man transformation of culture, ethics policy proposes the more mundane approach of re-conceiving ethics as the realizing of common democratic aspirations in more effective ways than have previously been the case.

But how rational are these democratic aspirations? Consider an observation from John Dewey (1859–1952), near the end of an extended essay on the need to unite ethics, policy, and rationality. "It is quite true," he admits, "that science cannot affect moral values, ends, rules, principles as these were once thought of and believed in." Then he continues:

"to say that there are no such things as moral facts because desires control forma-tion and valuation of ends is in truth but to point to desires and interests as them-selves moral facts requiring control by intelligence equipped with knowledge. Science through its physical technological consequences is now determining the relations which human beings, severally and in groups, sustain to one another. If it is incapable of developing moral techniques which will also determine these rela-tions, the split in modern culture goes so deep that not only democracy but all civi-lized values are doomed… A culture which permits science to destroy traditional values but which distrusts its power to create new ones is a culture which is destroy-ing itself." (Dewey 1939, pp. 117–118)

In other words, if we take policy thinking as an exemplar of the use of science and engineering in public affairs, we must admit that policy itself offers no insight or vision of the good independent of the needs and desires already manifest in the body politic. Its ability to help clarify, organize, and achieve the needs and desires as they are given deserves to be affirmed as a good. But is such a philosophical com-mitment to ethics policy sufficient to address the implicit nihilism that Dewey acknowledges? Or does it leave the door open to counter affirmations from alterna-tive sources such as tradition or religion?

4.4   Engineering, Technology, and Democratic Society

The brief examination of engineering and ethics in section two concluded with a question concerning whether engineers possess any knowledge that would justify their professional claim to special understandings of public safety, health, and welfare. The examination of ethics and policy in section three concluded with a related question concerning the rationality of democratic knowledge of the good. In more general conclusion, consider some complicating reflections on these two questions–reflections that provide modest support for the thesis that ethical rationality trumps technological rationality.

4.4.1 Engineering Knowledge

First, the issue of engineering knowledge: It is not clear that engineers as engineers know anything special about safety, health, or welfare in anything like the way physicians have special knowledge about the nature of health and lawyers about the structure of justice. The engineering education curriculum includes hefty doses of science, mathematics, engineering sciences (e.g., statics and thermodynamics), and design. But only in restricted ways do engineers learn about safety, health, and welfare (Mitcham 2009). The strongest exception is safety, but as the fact that there is a specialized discipline termed "safety engineering" suggests safety is not as integral to engineering as is sometimes proposed. By contrast, almost all medical school courses necessarily involve learning something about health; anatomy and physiology both include and explicate built-in notions about the proper structure and functioning of the human organism. Surely physicians know more about health–and welfare economists, perhaps, about welfare–than engineers know about these ends or ideals as such.

More specifically, on what possible basis are engineers more qualified than anyone else to understand or determine the safety, health, and welfare that should be associated with engineered structures, products, or processes? For instance, in Walter Vincenti's lucid analysis of *What Engineers Know and How They Know It* (1990) there is no indication that engineers qua engineers know anything special about safety, health, or welfare. Safety, health, and welfare are conspicuous by their absence.

Going further, the engineer philosopher Samuel Florman has argued explicitly with regard to safety that it would be crazy for engineers to determine "what criteria of safety should be observed in each problem" encountered. Artifacts can always "be made safer at greater cost, but absolute freedom from risk is an illusion." Levels of safety are "properly established not by well-intentioned engineers, but by legislators, bureaucrats, judges, and juries [and it] would be a poor policy indeed that relied upon the impulses of individual engineers" (Florman 1981, pp. 171 and 174). The most engineers can do is help clients and the public understand the relevant degrees of safety and then invite them to decide how safe is safe enough. Engineers qua engineers are no more qualified to make such judgments than anyone else; they legitimately participate in making such determinations, but only as users and citizens.

4.4.2 Policy Ends

Florman's argument implicates a second issue concerning ethics and policy. What is the basis for determining the good (or goods) that should serve as the end (or ends) in policy decision making? According to Dewey, ends should be determined by rational and intelligent democratic decision making. This is a contested proposition, but in the socio-historical period in which we live it remains the predominant view. Within such boundary conditions, then, what might be said about the rationality and ethical character of democratic decision making and participation processes?

The vision of engineering responsibility for more effective formulation and achievement of social policy goals, insofar as it limits citizen participation in decision making, functions as an implicit form of technoscientocracy. An engineer committed to the promotion of public safety, health, and welfare may make decisions about technical issues in an authoritarian manner at odds with democratic ideals, based on a strictly technical analysis and evaluation of the risks associated with some product or process. Recognition that technology often brings with it not only benefits but also costs and risks argues for granting all those affected some input into technical decisions. In result, at least two scholars have independently argued for a principle of "no innovation without representation" (Winner 1991; Goldman 1992).

One of the more provocative efforts to think through the participation principle in relation to engineering can be found in the collaboration of philosopher Mike Martin and engineer Roland Schinzinger. As their argument is stated in a widely adopted engineering ethics textbook, engineering should be seen as a form of "social experimentation." Engineering projects, which are increasingly integral to public affairs, are experiments insofar as they are undertaken in partial ignorance, outcomes are uncertain, and future engineering practice is modified by knowledge gained as a result. More crucially, these experiments impact users, consumers, and those societies in which the engineered structures, products, and processes are created and deployed.

Viewing engineering as an experiment on a societal scale places the focus where it should be: on the human beings affected by technology. For the experiment is performed on persons, not on inanimate objects. In this respect, albeit on a much larger scale, engineering closely parallels medical testing of new drugs or procedures on human subjects (Martin and Schinzinger 1983, pp. 59–60).

In consequence, "the problem of informed consent (…) should be the keystone in the interaction between engineers and the public" (Martin and Schinzinger 1983, p. 60). Stimulated by critical reflection on nuclear power and public policy, Kristin Shrader-Frechette has likewise argued that principles of free and informed consent should to be extended from biomedicine and be applied to the development of technology generally (1991 and 2002).

Just as medical research with human subjects or participants is moral only to the extent it respects the free and informed consent of the persons involved, so must engineering undertake to respect the autonomy of those it affects. Commitment to public safety, health, and welfare as a substantive ideal is replaced by the ideal understood in procedural terms. The basic form of safety, health, and welfare is not to be subjected to risks, deprivations, or harms to which one has not knowingly acceded, so that the practice of free and informed consent becomes the basic form of engineering ethics. Safety, health, and welfare are then publically determined by those affected through their free and informed participation.

There are nevertheless at least two key differences between informed consent in medicine and in engineering. Physicians and medical researchers possess substantive knowledge about the nature of health with which they can genuinely inform those who are exercising consent. Additionally, they are largely dealing with individuals. Neither of these conditions are met with regard to informed consent related to engineering projects. The character of safety is largely determined by the public involved and it is a public rather than individuals that is asked to exercise consent.

The difficulties of establishing appropriate protocols for practicing informed consent with regard to medical research and human subject participants are considerable; they can only increase when informed consent is raised to the societal level, as proposed by Martin and Schinzinger. Admitting the problematics of informed consent, they argue that at a minimum engineers have a responsibility to provide users and consumers "information about the practical risks and benefits of the process or product in terms they can understand" (Martin and Schinzinger 1983, p. 60). In a subsequent version of their argument (1989, p. 69) for the term "free and informed consent" they substitute "valid consent" (adapting from discussions in biomedical ethics literature). Can something more be said about the conditions for constructing valid consent, which could also be thought of as promoting a robust, democratic, rational, and intelligent determination of goods or ends?

Consider the question from two perspectives: one of technoscientists and another of democratic citizens. The first emphasizes the responsibilities of scientists and technologists to enhance (procedural) rationality, the second the responsibilities of democratic citizens to act (procedurally) rational. Together the ideal would be to construct a common technological and ethical rationality for a democratic establishment of a (substantive) good.

4.4.3   Responsibility of Technoscientists

With regard to the perspective of technoscientists or engineers one can derive useful suggestions from an analysis by political scientist Roger Pielke Jr. (2007), who distinguishes four idealized roles for a scientist or engineer who engages with publics: pure knowledge exponent, advocate, arbiter, and honest broker. This is an analysis that complements the more simple idea of an ethics code for the ethical exercise of technoscientific advice in the public sphere.

To illustrate the distinctions, imagine a politician or citizen seeking counsel regarding geoengineering responses to climate change. The pure knowledge exponent engineer responds like a detached bystander, spelling out in detail the various chemical and/or mechanical engineering processes that can sequester carbon. Politicians and citizens might well feel like they had inadvertently walked into a technical engineering class.

The issue advocate engineer, by contrast, acts like a salesperson and immediately argues for a bioengineering-related seeding of the ocean with iron to stimulate phytoplankton growth that would consume carbon dioxide. But the argument would be made with a peculiarly technical rhetoric that deploys information about the chemical composition of the iron, transport mechanisms, relation to phytoplankton blooms, and more. Politicians and citizens might well think they were standing at the booth demonstrating a proprietary innovation at an engineering trade show.

The arbiter engineer acts more like a hotel concierge. Steering a course between that of neutral bystander and advocate, such an engineer starts by asking what the politician or citizen wants from a geoengineering response: simplicity, low cost, safety, dramatic results, public acceptability, or what? Once informed that the aim is safety, the arbiter engineer would identify a matrix of options with associated low risk factors. The arbiter engineer engages with the public and communicates knowledge guided strongly by publicly expressed needs or interests. The concierge might on another occasion work as a medical doctor or psychologist counseling a patient.

Finally, the honest broker engineer reaffirms some modest distance from the immediate needs or interests of any inquirer in order to offer an expanded matrix of information about multiple geoengineering options and associated assessments in terms of simplicity, cost, safety, predictable outcomes, and more. The effect will often be to stimulate re-thinking on the part of inquirers, maybe a re-consideration of the needs or interests with which they may have been operating, even when they did not originally take the time to express them. The experience might be more analogous to a career fair than a single booth at a trade show.

Pielke's spectrum of alternative engagements between engineering and policy is certainly more adequate than any simply conceived, one-way, univocal engineering to policy model. It is also more robust than an ethics code. Although promoting the honest-broker role, Pielke is a pluralist insofar as he admits that any ideal type may be appropriate in the right context. Whatever ideal type is chosen, it just needs to be adopted with conscious recognition and transparent admission to any interlocutors. The advocate engineer, for instance, should say up front, "Let me tell you the technical reasons for adopting project X," making it clear that other engineers might well marshal knowledge in support of project Y. What is illegitimate, Pielke argues strongly, is stealth issue advocacy, which occurs when the advocate engineer fails either to recognize or to admit advocacy. It is even more illegitimate when advocates consciously hide or deny their advocacy.

Pielke and associates (Pielke et al. 2010) argue that a better path is to recognize the limits of engineering and to distance advice from interest-group politics while more robustly connecting it to specific policy alternatives. Research will not settle political and ethical disputes about the kind of world in which we wish to live. But engineers can connect their research with specific policies, once citizens or politicians have decided which outcomes to pursue. In this way, engineers provide an array of options that are clearly related to diverse policy goals. Rather than advocate a particular course of action, either openly or in disguise, engineers should work to help policymakers and the public understand which courses of action are consistent with our current–always fallible–technical knowledge about the world and our current–always revisable–visions of the good.

But there are inadequacies in this account of the engineer-politics interface as well. Although Pielke characterizes his four possible models for science or engineering to inform policy as ideal types, that of the so-called honest broker is impossibly ideal. One of the strongest findings in philosophically oriented science, technology, and society (STS) studies is that all knowledge production has what may be termed an engagement co-efficient. Engagement co-efficients can be strong or weak, but they cannot be avoided, simply because engineers are embodied, historical, and culturally situated persons. To think otherwise is to imagine engineering or technology as a neutral tool constructed on the basis of a view from nowhere. But engineering is constituted by a unique stance toward the human life-world and a particular, culturally influenced set of moral norms. To propose any technical input to policy is, at a minimum, effectively to affirm and promote the value of engineering and technology themselves.

Co-efficients of engagement can be further expected to include commitments not just to the value of engineering in general but to the values of particular branches of engineering, research programs, and more. The vicious interpretation of such unavoidable commitments is that all engineering is culturally biased or captive of special interests and no better than any other worldview, a position that promotes skepticism if not relativistic cynicism about appeals to engineering or technology. But one need not go this far, and in fact there are strong arguments for a qualified realist interpretation that grants what might be termed, adapting Philip Kitcher (2001), "well-ordered engineering" as an appropriately qualified but nevertheless privileged position in the political realm. No matter how high the wall separating engineering and politics, the relation between the two will involve dialogue and dialectic. It may not be possible to transcend the arbiter or concierge models, which nevertheless can be understood as contributing to enhancing public rationality (see also Collins 2014).

4.4.4 Responsibilities of Democratic Citizens

Turning from the responsibilities of technoscientists to those of democratic citizens, one can hypothesize a companion suite of ideal types. Pielke's four types are constructed from the point of view of the scientific experts. Another typology may be constructed from the perspective of those seeking scientific expertise, with such non-scientific principals are distinguished into those seeking agents, debate coaches, teachers, or proxies.

Again imagine politicians or citizens seeking counsel regarding geoengineering. Principals seeking engineers to act as agents (sometimes called "hired guns") want someone to scout out and/or help run a gauntlet of ignorance to realize their own goals. They do not want those goals questioned. Having made some decision in favor of or against some geoengineering project, these citizens are seeking expert witnesses to advance their cause.

In like manner, principals seeking engineering debate coaches want help in responding to anyone who might offer intellectual objections to a decision. Technoscientific assistants to politicians and corporate leaders often function in this manner. ("Tell me what science and engineering I can cite if objection X comes up.")

Principals seeking engineering teachers want to learn the science or engineering. They place themselves under the tutelage of technoscientists in order not to become scientists or engineers themselves but to acquire the knowledge of, for instance, science journalists or simply well informed citizens.

Finally, at the opposite extreme are principals who want technoscientific proxies to whom they can delegate decision making, trusting they share or will respect the principal's basic values and commitments. This readily occurs when medical patients are so overwhelmed with information about an illness that they feel unable to decide between alternative treatments offered by an attending physician. ("Doctor, you decide. You know better than me. I cannot think about it anymore.") It also occurs when congress or a public charges technical agencies with delegated rulemaking.

As is obvious, principals have needs and interests that can bias their use of technoscience just as scientists and engineers have needs and interests that can bias their inputs to policy. In any technoscience for policy both deserve to be acknowledged and should enter the dialectic or dialogue. Additionally, however, the ideal stance of principals seeking teachers–that is, principals who are in the first instance good listeners and learners–is arguably the one most likely to enhance democratic rationality.

To illustrate this point, it is possible to distinguish first-, second-, and n-order goods in well-designed structures, products, processes, and systems. Especially with technologies, first-order technical goods are constituted by the functioning of the object itself. This functioning can be strongly or weakly coupled to second- and n-order goods of a more public character. But the second-order goods are more issues of public than of engineering judgment. This fact offers another way to interpret the phrase "public safety, health, and welfare." It is not so much that engineers have a responsibility to determine and protect public safety, health, and welfare as they see it, but that they have a responsibility to protect what the public determines as its vision of safety, health, or welfare. The more remote the order of goods and the more weakly coupled the relation between first- and n-ordered goods, the more this will be the case.

Engineered public goods in many instances are easily determined by the public that benefits directly from engineered buildings, consumer products, production processes, and transport or communication systems. Engineers thus have a corresponding duty to take into account how the public may indeed benefit, while being granted a certain degree of autonomy, not so much to converse among themselves (like scientists) as to converse with the public. This should include especially an ability to alert the public about situations that pose dangers or risks to use that may be obscured by short-term convenience or business interest.

Insofar as there is a tight coupling of first- and n-order goods, engineering also properly plays a more robust and direct role in decision making with regard to those projects it is assigned by political or policy decision. When President John F. Kennedy in 1961 committed to a Moon landing in less than ten years he could do so only after consulting with engineers about the feasibility of such a project. He (or his advisers) had to seek teachers of the technically feasible and be willing to listen and learn.

Similar consultation is required with regard to any technological project to be undertaken as a result of political or commercial decision making. When engineers are instead forced to behave as agents or debate coaches in the service of unrealistic goals, disaster often ensues, as was the case in the Soviet Union when Stalin enrolled engineers in impossible projects (Graham 1993).

Turning back to the responsibilities of engineers: Engineers themselves, independent of being consulted about externally conceived projects are regularly putting proposals before politicians, the public, and venture capitalists for what they imagine as new, reasonable engineering undertakings. In the process, they have obligations to bracket their technical enthusiasms in favor of thinking beyond the technical to consider as honestly as possible the potential for real public benefit. Indeed, commercial employers often admonish engineers to take on the perspective of management and try to anticipate the economic implications of their work. This is another version of the duty *plus respicere*, to take more into account: a general, abstract statement of the conditions for procedural rationality under conditions created by technology.

4.5   Coda

In an insightful reflection on *The Techno-Human Condition* (2011), Braden Allenby and Daniel Sarewitz offer another take on this tension between first- and n-order goods. They distinguish three levels of cause-and-effect relationship between engineering and the world. Level I occurs when a specific device is engineered to achieve a clearly defined function: "a vaccine prevents a particular disease, or a well-design manufacturing process eliminates the use of toxic chemicals." Here one can be confident of the results because a policy good (health) and engineered instrumental means are simple and direct. First- and second-order goods are tightly coupled.

In Level II relationships, however, engineering becomes part of "a networked social and cultural phenomenon [functioning] in a broader context that can be complicated, messy, and far less predictable or understandable." Examples include transport and communication systems. At Level II "acting to achieve a particular intended outcome is often difficult because the internal system behavior is too complicate to predict." First- and n-order goods are less tightly coupled.

The first two levels are relatively familiar and engineers can make reasonable efforts to assess the outcomes, intended and unintended, of their actions into systems. At Level III, however, the system has become so large and complex that its boundaries are difficult to determine; it has become a "complex, constantly changing and adapting system in which human, built, and natural elements interact in ways that produce emergent behaviors which may be difficult to perceive, much less understand and manage." At this level, it is increasingly difficult to be sure about the relation between first- and n-order goods. Additionally, our experience of the techno-life-world, according to Allenby and Sarewitz, is that "We inhabit Level III, but we act as if we live on Level II, and we work with Level I tools" (Allenby and Sarewitz 2011, pp. 63 and 161).

What does this recognition imply? The Allenby-Sarewitz insight is an ethical one. To summarize in the briefest possible terms: The rationality of ethics concerns how to allow ends to emerge. Rationality of technology concerns how best to achieve the ends once they have been posited. Ethical rationality thus trumps technological rationality. The former emphasizes the ethical responsibilities of citizens to act rationally and with intelligence by incorporating science and engineering into deliberation with regard to ends, the latter the responsibilities of scientists and engineers to act rationally and with intelligence in advising democratic citizens.

Acknowledgments This paper draws heavily on and adapts extensively from collaborative work with Adam Briggle and Erik Fisher. See especially Adam Briggle and Carl Mitcham 2012. *Ethics and Science: An Introduction*. Cambridge: Cambridge University Press, particularly chapter 12; and Carl Mitcham and Erik Fisher. 2012. "Ethics and Policy." In Ruth Chadwick (ed.), *Encyclopedia of Applied Ethics*, 2nd edition, vol. 2. San Diego: Academic Press, pp. 165–172.

Thanks also to Wenceslao J. Gonzalez for encouragement in developing this effort in critical reflection and for his insightful comments on earlier versions of the argument. His own work, especially on issues related to rationality and values, deserves to be more fully considered than has been the case here; it will undoubtedly be important to future discussion and analysis.

References

Alexander, J.K. 2008. *The mantra of efficiency: From waterwheel to social control*. Baltimore: Johns Hopkins University Press.

Allenby, B., and D. Sarewitz. 2011. *The techno-human condition*. Cambridge, MA: The MIT Press.

Blockley, D. 2012. *Engineering: A very short introduction*. Oxford: Oxford University Press.

Briggle, A., and C. Mitcham. 2005. Bioethics committees and commissions. In *Encyclopedia of science, technology, and ethics*, vol. 1, ed. C. Mitcham, 202–207. Detroit: Macmillan Reference.

Briggle, A., and C. Mitcham. 2012. *Ethics and science: An introduction*. Cambridge: Cambridge University Press.

Collins, H. 2014. *Are we all scientific experts now?* Cambridge: Polity.

Dewey, J. 1939. *Freedom and culture*. New York: G. P. Putnam's.

Diamond, J. 2005. *Collapse: How societies choose to fail or succeed*. New York: Viking.

Dupree, A.H. 1957. *Science in the federal government: A history of policies and activities to 1940*. Cambridge, MA: Harvard University Press.

Environmental Protection Agency (EPA). 1999. *Twenty-five years of the safe drinking water act: History and trends* (EPA Report 816-R-99-007.) Washington, DC: Environmental Protection Agency.

Fisher, E. 2005. Lessons learned from the Ethical, Legal and Societal Implications program (ELSI): Planning societal implications research for the National Nanotechnology Program. *Technology in Society* 27(3): 321–328.

Fisher, E., and R.L. Mahajan. 2006. Contradictory intent? US federal legislation on integrating societal concerns into nanotechnology research and development. *Science and Public Policy* 33(1): 5–16.

Florman, S. 1981. *Blaming technology: The irrational search for scapegoats*. New York: St. Martin's Press.

Frankena, W.K. 1983. Concepts of rational action in the history of ethics. *Social Theory and Practice* 9(2/3): 165–197.

Galilei, G. 1638. [*Discourses on the*] *Two New Sciences*. Trans. S. Drake. Madison: University of Wisconsin Press, 1974; 2nd edition, 1989, and 2000 Toronto: Wall and Emerson.

Gewirth, A. 1992. Rationality vs. Reasonableness. In *Encyclopedia of ethics*, ed. L. C. Becker and Ch. B. Becker, vol. 2, 1069–1070. New York: Garland. Expanded version in 2nd edition, vol. 2, 1451–1454 (New York: Routledge, 2001).

Goldman, S.L. 1992. No innovation without representation: Action in a democratic society. In *New worlds, new technologies, new issues*, ed. S.H. Cutcliffe, S.L. Goldman, M. Medina, and J. Sanmartín, 148–160. Bethlehem: Lehigh University Press.

Graham, L. 1993. *The ghost of the executed engineer: Technology and the fall of the Soviet Union*. Cambridge, MA: Harvard University Press.

Hollander, R.D., and N.H. Steneck. 1990. Science and engineering-related ethics and values studies: Characteristics of an emerging field of research. *Science, Technology, and Human Values* 15(1): 88–104.

Hume, D. 1751. *An enquiry concerning the principles of morals*. A. Millar, London. Reprinted with Tom L. Beauchamp as editor (The Clarendon Edition of the Works of David Hume). Oxford: Oxford University Press, 1998.

Institution of Civil Engineers (ICE). 1818–1823. *Minutes of the proceedings of the Institution of Civil Engineers*, 1. Unpublished archives, London: Institution of Civil Engineers.

Kahneman, D. 2011. *Thinking, fast and slow*. New York: Farrar, Strauss and Giroux.

Kitcher, Ph. 2001. *Science, truth, and democracy*. New York: Oxford University Press.

Langfelder, E., and E. Juengst. 1993. Profile of Ethical, Legal and Social Implications (ELSI) Program, National Center for Human Genome Research. *Politics and the Life Sciences* 12: 273–277.

Lasswell, H.D. 1951. The policy orientation. In *The policy sciences: Recent developments in scope and method*, ed. D. Lerner and H.D. Lasswell, 3–15. Stanford: Stanford University Press.

Martin, M. W. and R. Schinzinger. 1983. *Ethics in engineering*. New York: McGraw-Hill. Second edition, 1989. (Fourth edition, 2005).

McCarthy, N. 2009. *Engineering: A beginner's guide*. Oxford: Oneworld.

Mitcham, C. 1994. *Thinking through technology: The path from engineering to philosophy*. Chicago: University of Chicago Press.

Mitcham, C. 2009. A philosophical inadequacy of engineering. *The Monist* 92(3): 339–356.

Mitcham, C. 2011. Technology and the burden of responsibility. In *Values and ethics for the 21st century*, ed. F. Gonzalez et al., 169–197. Madrid: BBVA. Spanish version: "La Tecnología y el peso de la responsabilidad." In Gonzalez, F. et al. 2011. *Valores y ética para el Siglo XXI*. Madrid: BBVA, pp. 177–207

Mitcham, C., and E. Fisher. 2012. Ethics and policy. In *Encyclopedia of applied ethics*, vol. 2, 2nd ed, ed. R. Chadwick, 165–172. San Diego: Academic.

Mitcham, C., and E. Schatzberg. 2009. Defining technology and the engineering sciences. In *Philosophy of technology and engineering sciences*, Handbook of the philosophy of science, vol. 9, ed. A. Meijers, 27–64. Amsterdam: Elsevier.

Moser, P. K. 1996. "Rationality." In *Encyclopedia of Philosophy Supplement*, 488–490. New York: Macmillan. Included unchanged in *Encyclopedia of Philosophy*, 2nd ed., 2005. Detroit: Macmillan Reference, pp. 253–255.

NIH guide for grants and contracts. 1989. 18(45).

Pielke Jr., R.A. 2007. *The honest broker: Making sense of science in policy and politics*. Cambridge: Cambridge University Press.

Pielke Jr., R.A., D. Sarewitz, and L. Dilling. 2010. *Usable science: A handbook for science policy decision makers*. Boulder: Center for Science and Technology Policy Research.

Shrader-Frechette, K. 1991. *Risk and rationality: Philosophical foundations for populist reform*. Berkeley: University of California Press.

Shrader-Frechette, K. 2002. *Environmental justice: Creating equality, reclaiming democracy*. New York: Oxford University Press.

Simon, H. 1969. *The sciences of the artificial*. Cambridge, MA: The MIT Press. (Third edition, 1996.)

Simon, H. 1983. *Reason in human affairs*. Stanford: Stanford University Press.

Tredgold, Th. 1828. Development of a Civil Engineer. *Minutes of the Proceedings of the Institution of Civil Engineers* 2, Meeting of Council, January 4, 1828: 20–23.

Veblen, Th. 1921. *The engineers and the price system*. New York: B. W. Huebsch.

Vincenti, W. 1990. *What engineers know and how they know it*. Baltimore: Johns Hopkins University Press.

Weirich, P. 2008. Rationality. In *International encyclopedia of the social sciences*, vol. 7, 2nd ed, ed. W.A. Darity Jr., 79–82. Detroit: Macmillan Library Reference.

Winner, L. 1991. Artifacts/ideas and political culture. *Whole Earth Review* 73: 18–24.

Chapter 5
Knowledge and Moral Responsibility for *Online* Technologies

Juan Bautista Bengoetxea

The first IBM personal computer was unveiled in 1981 and served as the starting point for several debates about the relationship between humans and machines. Some international journals substituted human image for those of computers on their covers,[1] beginning so a new way of thinking of possible negative consequences of new technologies. Currently we are not afraid of personal computers, though perhaps we are of the *significance* they might have in our lives. As far as we are cyborgs or extended minds,[2] we should be aware of the relationships we have with those technologies that are becoming part of us, and try both to avoid any sort of dependency on them and to control them somehow. The development of information and communication technologies (ICTs) and their networks is helping us clarify some problems previously hidden behind philosophical and academic debates. In this text, I shall focus on some particular questions of epistemology and ethics related to those debates.

The Problem Information and knowledge are basically generated in practical contexts. One of special interest is the *online* activity context. Knowledge is a social outcome and it is no doubt linked to ethics, norms, and values (Brandom 1994, p. 170). In the case of virtual knowledge, its ethical and social nature takes firmly root in its practices. It is therefore important to articulate the shift from the traditional notions of knowledge and justification—laden with views based upon concepts such as *evidence* and *perceptual access to data* in order to get reliable knowledge— to a new inquiry on online knowledge and information practices according to recon-

---

[1] *Time*, January 1983.

[2] Here I take Clark's claim that we are cyborgs (2003, p. 5). Accordingly, what is special in the human brain and better explains the main features of our intelligence is just the capacity of our brain to be *situated* in complex relationships with respect to non-biological "constructs."

J.B. Bengoetxea (✉)
Department of Philosophy and Social Work, University of Balearic Islands, Crta.
Valldemossa, km. 7,5, Palma de Mallorca 07.122, Spain
e-mail: juanbautista.bengoechea@uib.cat

© Springer International Publishing Switzerland 2015   89
W.J. Gonzalez (ed.), *New Perspectives on Technology, Values, and Ethics*, Boston Studies in the Philosophy and History of Science 315,
DOI 10.1007/978-3-319-21870-0_5

structed criteria, which allow us to talk in terms of knowledge and justification—*genuine* as well—in these cases. Obviously, this task requires a reflection on the very notion of online community, as well as to peruse how to implement, if possible, ethical and normative codes in those communities. The present text is also dedicated to these topics.

The View The text is structured as follows. First, I examine some particular features of the Kantian and Hegelian philosophies of both the social role of practices and their relationship to ethics. This way I try to conceptually articulate the basis for a discussion of norms and values in online practices. As a result, an outline similar to Nagel's "fragmentation of values" (1979) is proposed. The second and third sections are the core of my proposal: the former is a discussion of the possibility to create a philosophical realm focused on virtual epistemic practices that could allow us to *reformulate* the notions of knowledge and justification. The third section focuses on the behaviour of people involved in online activities of information and knowledge. As a consequence, I claim that it is necessary to reconstruct many traditional and customary values, norms and ethical codes of non-online contexts. The conclusion is a brief plea for the usefulness of ethical codes in epistemic online practices.

5.1   Practices, Principles, and Virtuality

The discussion on how to establish ethical norms in new contexts such as online ones can be set on two philosophical bases: by reference to *principles* and by reference to *practices* or traditions. It is uneasy to say what characterizes the ethical and what the unethical, but at least we can recognize that social individuals share moral traditions and several practices. Among these practices, the online activity is a quite relevant kind that generates epistemic disagreement and deserves attention.

What is the relevance of practices in our ethical judgments about the use of, say, Internet? How do those practices affect the user? Technology used in scientific and cognoscitive areas, in data-search tasks, in the creation of new phenomena and facts, and in search and information storage has several important cognoscitive advantages. For example, we could say that reliable epistemic practices that improve human intellectual capacities by means of computerized devices and electronic networks also "reinforce" epistemically users and researchers. I.e., users may improve their reasoning and decision capacities by using computer devices and communication technologies. Later I shall examine some epistemic norms that usually are taken into account in online context ethics.

However, now I am interested in presenting an attempt to overcome the above mentioned dichotomy between practices and principles. Do these two notions represent two different accounts of social activities? Which of them would be more efficient to analyse epistemically and ethically online activities? Let us summarize the basic point of both in two brief statements:

> *Principles view* [PriVi]: There are aprioristic abstract principles that can be applied to the domain of online knowledge activities.
>
> *Practices view* [PraVi]: First, it is necessary to examine real practices and their rules and then, if some principles underlying them have been found, it is concluded that they are principles implicit in practices, but not aprioristic ones.

As we know, there are wrong practices to which we should not be deferential. They ought not to be accepted as an empirical basis from which to derive rules. Instead, it would be more efficient to generate standards of use (*norms*) in virtue of which to distinguish cases of right conduct from those of wrong conduct, without forgetting online practices for attempting to improve those norms. In addition, we should emphasize the extreme importance that online practices have in the formation of a framework for epistemology and ethics to this day.

Social practices make reference to those behaviour standards that may be recognized in human communities. Along with notions such as custom, convention, institution, tradition, or norm, social practices belong to what Kant called "anthropology." Although a practice is not necessarily governed by rules, we often conceive of it that way. While several practices such as chess have explicit rules that indicate what is right and what is wrong, most of practices do not have this kind of rules. Hence, the term "practice" also refers to ways, not defined by rules, of making things. It is precisely in this sense that we can understand virtual practices.

Kant's claim was plain: we cannot derive ethics from anthropology. He did not give any relevant role to human behaviour in the issue of ethical necessity (Tunick 1998, p. 11). One thing is to *recognize* that something is right (or wrong), and quite another to know if it is *essentially* so, regardless of who performs such recognition—one individual, two, a nation, a law, a costume, a convention, a shared idea, or a norm. It was Hegel who opposed this theoretical view when he focused on practices from a radically different point of view, namely from his *PraVi*. Remember that the Kantian moral imperative forces us to have a prior duty to any kind of experience, so that morality holds an aprioristic nature for that imperative. Thus, Kant adheres to *PriVi*. Hegel's philosophy, however, rejects the idea of a non-empirical ethics (Hegel 1821, §1). He is not rejecting all principles, but rather conceives of principles as something immanent in our practices, being the latter prior to the former. Reason must be used in an a posteriori way.[3]

According to *PriVi*, contents of ethics are not determined by observing social practices, but rather by establishing an abstract categorical imperative that leads us to make moral judgments that avoid any appeal to norms and social practices. In this regard, this kind of practices is not, and cannot be, a source of authority for moral

---

[3] To this respect, see the interesting edition of Priest (1987).

will, to the extent that morality requirements are not determined by convention or practice.[4] In addition, Hegel insists on the fact that, in order to establish what to do, we have to analyse social practices. He thus deeply criticizes the categorical imperative because this is a principle of merely abstract universality whose determination is an identity without content; i.e., it is "the indeterminate" (Hegel 1821, §1).

According to *PraVi*, Kant is unable to give up his formal point of view, just designed to impose, under some moral maxims, a mere *non-contradiction* requirement. Hegel instead claims that there are well-established actual principles that determine how to act. They are principles that we find out in an ethical life, when we "*move toward* the concept of ethics" and appeal to a moral guide not by means of a formal non-contradiction principle, but rather through immanent principles belonging to our shared practices and understanding. Thus, practices and institutions of an ethical life provide us the standards for right behaviour, and just when our consciousness— subjective standards of action—fits those objective standards, we say that we are acting correctly.

Here I am going to invoke Thomas Nagel's (1979, originally published in 1977) proposal for combining two types of perspectives like those exemplified by *PriVi* and *PraVi*, since I deem it is sufficient to project a more complete epistemic-ethical analysis of online contexts. Nagel claims that the disparity of value fragmentation, on the one hand, and the uniformity of a decision, on the other, precipitates many problems that take the form of practical conflicts. According to him, there are five fundamental value types that create basic conflicts (Nagel 1979, pp. 129 and ff): specific obligations, general rights, utility, perfectionist values, and project commitments. Among the several proposals to build up hierarchies on these types, so far none has been successful due to the lack of 'uniqueness' in the source of value.

Nagel's strategy does not seek a single general theory that is useful to know how to decide what to do in a right way (Nagel 1979, p. 135). The progress made in the processes of justification and systematic criticism of beliefs does not stem from general principles of reasoning, but rather from the particularized analysis of specific doxastic contexts. However, this kind of fragmentation of beliefs is not at stake and therefore it is difficult to recognize any fragmentation applied to decision making.

Hence, the most convenient solution seems to be a compromise that combines systematic outcomes, when they are applicable, with less systematic judgments that close the possible fissures that can appear among the former. Obviously, this point

---

[4] There is a realm that Kant does accept to be that of practices, namely the realm of *law*. By appealing to conventional attitudes in order to determine penal laws, Kant is not contradicting his view that morality is determined without appealing to practice or convention, since for him the spheres of morality and law must be distinguished. By reflecting on what laws need, Kant thinks that it is right to take consequences into account, although it is not the case of moral. Kant's moral theory does not appeal to norms and social practices because his theory is a deontological one. If ethics must be a categorical and non-hypothetical discipline, then the determination of what ethics needs cannot depend upon sanctions and awards imposed by practice. And since Kant believes that legal theory, in opposition to moral theory, is pragmatic and consequentialist, he is consistent when is appealing to social practices in his doctrine of right.

of view needs to be completed with a developed account of the nature of decisions, as it is intended in this text. I would say that the Nagelian compromise is a rational approach beyond the *practices/principles* dispute; i.e., that it is a method of analysis of practical problems that might guide us on the way forward in order to apply valuable principles to particular cases. Hence, it is a conviction underlying this project that ethics itself is not a decision procedure, but rather a resource that can advise us in decision making tasks.

5.2 Epistemology: Information and Knowledge

Although the analysis of virtual practices has usually taken an axiological path,[5] I would wish to emphasize the value and relevance that epistemological features have for our reflection on the significance and influence of online practices in philosophy.

In the Internet case, the starting key-question can be established in the following terms: Does Internet change in some way the nature of knowledge and justification in online practices? What does Internet add to the "traditional" notions of knowledge and justification? It seems to me that online activities force us to conceptually re-examine the perpetual problem of knowledge and justification.[6] Particularly, Internet has challenged the classic philosophical notion of *agency*, both individual and communitarian. According to this view of the cognitive agents, it is just our capacity to control world events in perceptive ways—by direct access—what makes it possible for us to refer to some reality. We can identify who are cognoscent agents and who are not, assess them, and even give and request them reasons about assertions that they express or actions they do. But the case is that Internet has made more complicate this process of agency identification.

5.2.1 Cognitive Agency in Online Contexts

The reformulation of the nature of knowledge and justification in online contexts is a task that has brought about a host of replies. Among them I would underline Goldman's (1992, p. 179 and ff) *veritistic-social* response, which can help us obtain

---

[5] Axiological questions, focused on topics such as privacity, property or freedom of speech, have embrace almost all the philosophical debate on the virtual, but often leaving aside epistemology (see Dreyfus 2001; Johnson 1994).

[6] Here I'm not going to compare epistemological theories of knowledge with those of justification, but anyway I recommend Langsam's paper (2008) for an elucidation of the state of the art in the context of the debate between externalists and internalists.

a more accurate view of the roles of knowledge (or belief) and justification in online epistemic practices.[7]

Goldman proposes to divide epistemology into two branches: *individual* and *social*. Both would aim to identify and assess processes, methods, and practices in terms of their contribution to the production of true beliefs. While the individual branch identifies and assesses the psychological processes of an epistemic subject, the social does the same in the case of social processes through which epistemic subjects interact with other agents that causally influence on their beliefs (Goldman 1992, p. 181). In this regard, communicative actions between agents and institutional structures, which frame—or direct—those actions, would be paradigmatic of the kind of social-epistemic practices that we should examine in an epistemological veritist-social account.

According to Goldman, in both ordinary life and specialized knowledge we prefer to have true beliefs rather than false ones—or no one: uncertainty. This is what Goldman calls the "*veritistic* value."[8] There are several social and technological practices that feed this value, for example the discursive (information, argumentation) and the informational-technological practices. Goldman's account, accordingly, tries to be *normative*. It is intended to assess actual practices in terms of the impact they have on beliefs and on the decision on whether these are true or not. That is why it is important to realize that the role agents play in online practices is basically social. Even though every user is individualized, the Internet online activity depends basically on its links to other users. The Web needs many contact points in order to work out efficiently, being very significant the fact that not every social epistemic assessment criterion fits online practices assessments. Therefore, Goldman (1992, p. 183) prefers to decrease the number of social assessment targets to three: individual's beliefs, social belief profiles, and social practices, procedures and institutions. This last goal is the closest to an epistemology of online practices.

5.2.2   Ethics and Online Social Knowledge

Traditionally, knowledge has been conceived as *justified true belief.*[9] But if justification would not work, then what Gettier (1963) called "knowledge by chance" (or something akin to knowledge but that is actually not) might emerge. Gettier showed then the possibility that epistemic chance is not completely eliminable, especially because the traditional definition is quite weak in front of Zagzebski's (1999)

---

[7] Another interesting reply that I do not analyze here is Ihde's phenomenological view (Ihde 1990).

[8] Goldman proposes his *veritism*—understood as a social epistemic assessment—as an alternative option to other evaluative views, particularly to *consensualism* and *expertism* (see Goldman 1992, pp. 186–189).

[9] S knows that *p* iff (i) *p* es true, (ii) S believes that *p*, and (iii) S is justified to believe that *p*.

"double chance" argument.[10] For knowledge to be genuine, Goldman (1967) proposes a kind of *infalibilism* (1967) that requires the knower to be in a causal connection with the evidence that supports the belief and its articulation.[11] If there is not such a connection, believing *p* is not justified. In the online knowledge, hence, we find out something like Gettier's counterexamples, whose goal was to prove that the traditional definition of knowledge was not sufficient. The difficulty for users of online information systems has to do with the justification of propositions or existing data. It seems that this kind of users depends at the most upon the confidence in some authorities, who in turn allow them to obtain the minimum credibility in front of the difficulty (maybe impossibility) of establishing the causal connections to the evidences supporting their inferences and knowledge assertions.

In order to understand this dissatisfaction in online cases, we have to look at to the social dimension of knowledge. The reluctance to accept a non-traditional epistemology derives, at least partially, from the fact of not recognizing the cooperative nature of cognitive communities. Inertia in traditional analyses of knowledge has led to recognise only subjective elements—in the form of, say, "mind," "self," "person," or "subject"—as a source of possible epistemic certainty that could avoid the vulnerability with which scepticism challenges us. But we know that knowledge processes are complex and delicate networks of reliability—and reliance, especially in *online* cases.

Therefore, we can find two essential features that meet at this point, one epistemological and one ethical. Any examination of online epistemic practices requires an ethical scrutiny of actions of those practices. It seems quite clear that when we use computational means for collaborative research, we have the moral responsibility of both watching out for them and assessing whether they are trustworthy or not. This is so because computational means are different with respect to *offline* contexts, in which epistemic and moral responsibilities have a clearer delimited place. In the latter case, the *agent* (the knower) is responsible just for not having reached the intended purposes. In online contexts, however, responsibilities of knowing and acting have begun to be diluted.

The previous outcome is partially a result of the surface appearance of online epistemic and ethical practices. It may be unnoticed that communication among persons as well as reading and writing in Internet bring about something especially new and relevant from a philosophical point of view. But this assessment would be wrong unless we understand that the novelty of this new context changes the epistemic-ethical background. Some problems with the same old name—but with the addition of an "e"—might become something quite different now. This fact has been reflected in disputes over Internet values and norms and, accordingly, in those values and norms inherent to hypertexts and their contents that did not exist previously,

---

[10] See Zagzebski (1999), p. 100. In "double chance" cases, the subject has *evidences* that allow her to believe that *p*, despite evidences are not in fact related to the truth of *p*. However, chance makes *p* to be true.

[11] According to Goldman, *perception, memory,* a *causal chain,* or different combinations of these three items are good candidates to be a causal connection.

but that influence our notions of both knowledge and morality. I classify them in five sets summarized under the following labels: *information integrity* (accurateness and reliability), *epistemic authority*, *agency capacity*, *fertility*, and *efficiency* (see Goldman 1992, p. 195; Weckert 2000, pp. 48 and ff; and Thagard 1997). Hence, agency will be identified in virtue of how we conceive of these sets selected according to an empirical inquiry of actual online practices.

Integrity is understood here as the attitude that aims at accurateness and reliability of what is made and poured in information and knowledge networks. We must not forget that beyond the safety and confidentiality of Internet data it is important to recognize the ownership of ideas belonging to others. Reliability—measured by the ratio of truths to total number of beliefs fostered by a practice—seems to be one of the most basic questions about Internet, given the special capability of data spreading in this environment. Not only personal information but also publications and knowledge products are on hand to anyone. It is true that search engines can supply plenty of data about many *original* sources of knowledge, but also about *mediators*, and these are not always a reliable source with respect to the originals.

In the first case—access to experimental data, papers, texts, or graphics—, the problem source may be the agent who makes use of such information. It is supposed that she is honest in face of cases of plagiarism. In the second case, however, reliability—or lack of it—is inherent to the very information we could obtain. Who can ensure that the nth source of information about Locke's works has not wrongly paraphrased the n-1th and plagiarized the n-2th? If information is not reliable, those who use it are those who may suffer the consequences in the subsequent process of justification of their writings as putative knowledge. It should be emphasized that the great amount of online information makes the process of reliability identification much more difficult. Hence it seems to me that *authority* should play here a similar role than in traditional contexts.

Therefore, the quantitative factor makes the issue more complex in a sensitive way, and to think that non-qualitative or non-conceptual problems are not really significant for the philosopher would probably be a serious mistake. In fact, they are not "soft problems" (see Chalmers 2002, p. 247). The emergency of qualitative features from the complexity of the quantitative has usually been a *pattern* across the history of knowledge and human practices. Today it is quite easy for anyone to publish "something" in the virtual space. There are many information sources and it is easy to spread data thru cyberspace in search of a receiver. Still, since there are not many philosophical online publications that satisfy traditional standards of quality, the online potential for inaccurateness is higher and options for, say, libels are quite high. All of this means that anyone could freely publish a product—probably of low quality—without taking field authorities into account.

The capacity of agency is identified as the ability someone has to help other agents find out good answers to topics of interest for the latter. Thagard describes six elements of this capacity: (i) Online and hypermedia techniques: these lead agents to find answers to their questions in a more efficient way than the traditional one, and also supply techniques of representation of broad scope, inaccessible otherwise. (ii) Hypertexts: these make easier to trace new sources of information previously unknown. (iii) Digital data-bases, which can be fastly and easily found out. (iv) Electronic mail and newsgroups: these are potential sources of several capacities. It is also interesting to discriminate between groups with a moderator and those with none: the former provide relevant information, sometimes knowledge, to the agency; the latter groups supply, instead, slag-information.[12] (v) Furthermore, it is too a fact of matter that in Internet is possible to discover ready-to-use software, sometimes free, which allows the agent for better working conditions. Last (vi), pre-print electronic files increase an agent's capacity to detect answers to her questions.[13]

Efficiency incorporates *costs*. Whenever we seek answers to new and complex questions, we need some resources that help us acquire them. More efficient practices are those that promote acquisition at lower cost. Internet is well suited to efficiency standards: e-mail or electronic files, say, require lower costs than their paper correlates. All these ingredients are implicit in online epistemic practices, though making them explicit is not always an easy task. Notwithstanding, I think it may be convenient to present some reflections about this, focused on moral aspects closely linked to epistemic ones.

5.3 Ethics: Norms and Codes for Virtuality

Online practices may be right or wrong (*malpractices*). Their open nature allows inadequate information — false, bad mannered, unclear, biased — to be easily exposed and to create a place for irresponsibility. Before the end of the twentieth century, we have seen some initial signs of cyber-wars, false identity, systematically manipulated information, ideologized discourse transmission, and even questionable forms of "hacker activism." It may be a task for philosophers to analyse possible *moral* assessments of online behaviours with respect to the development and improvement of *knowledge* and its justification. Since applied ethics has already shown that starting from a deontological-theoretical, teleological, or strictly normative toolkit about ethical virtues is a wrong way of conceiving what applied ethics actually is, here we propose an alternative akin to Nagel's account.

5.3.1 Morality and Online Knowledge

Reflection on moral aspects of online practices must take into account at least four kinds of problems: jurisdictional and projective, applicative, problems of individuation, and problems of moral ignorance.[14] The four kinds can be subsumed into a

---

[12] Thagard invites us to compare a moderated newsgroup (http://sci.physics.research) with a non-moderated one (http://sci.physics) in order to see differences in quality and reliability.

[13] "Physics" file and "cognitive science" file (http://cogprints.soton.ac.uk/) are two good examples.

[14] See van den Hoven (2000), pp. 129 and ff.

fourfold classification ordered in a progressive increase of difficulty (van den Hoven 2000, pp. 133 and f):

(i) There are issues merely *related to* online environments. In order for these issues to emerge, an online environment is neither necessary nor sufficient,[15] since those are questions that also emerge in traditional offline environments. If some institution, say, an online University with which we contact by means of computational technology, does not meet the required information it "sells" in Internet, the moral issue so generated is that of "misappropriation," "cheat" or "negligence," but not a case of "online fraud." This kind of issues is at the bottom in online practices.

(ii) There are issues that *depend upon* online activities. Online technologies are necessary but not sufficient for them to create a real moral problem. Two cases are network information transfers and some forms of hacking access data and expansion of viruses. All those processes need computational networks and security technologies.

(iii) Likewise, there are some issues that are *determined by* online practices. In this case, moral problems arise just if computational applications start to work. In this sense, online technologies are a sufficient but not necessary condition for moral problems to emerge. A good example is the "fair access," or else the "moral responsibility with respect to the quality of online information." It is not necessary, since the same "quality" problem could emerge in printed books and in non-virtual information sources.

Finally, (iv) there exist *specific* questions that *only belong to* online practices. Online practices are necessary and sufficient in order for this kind of moral problem to arise. Hence, it is a kind that does not emerge in other environments. Those are problems linked to emergent values (Steinhart 1999), to the new AI, or to artificial life in the form of autonomous agents (*softbots*).

These reflections add to the accepted double belief that we all intend to carry out practical goals in social contexts and that we do it in such a way that the abovementioned necessities are especially relevant. In this regard, online actions and behavior rules related to information and situated in some of the four kinds of issues are subjected to a moral assessment just because they are crucial for community agents to be able to develop knowledge (Schmidtz and Goodin 1998).

5.3.2   *Moral Values, Knowledge, and Epistemic Responsibility*

Nagel's value topology supplies an interesting framework to assess online individual decisions and actions from a moral point of view. It describes several kinds of values that can be understood as a source of incompatible moral advice, so that a combined classification would reflect more accurately both the fact that human

---

[15] Two examples are (non-) veracity in advertisement and fraud in electronic trade.

beings usually focus on decisions from different perspectives—personal, social, human-global—and the fact that assessments are diverse: some are based on fundamental values oriented to results (*usefulness*), some are abstracted from action consequences (*duty* and *obligation*), and other are personal and based on individual agents.

The *utility* type-value proposed by Nagel points in a decisive manner to the epistemic utility subgroup. This kind of value takes into account an agent's positive and negative effects of his actions on the well-being of an individual. The latter may be interpreted in terms of money, flourishing, or fullfilled preferences, but also in terms of fair partition—material or abstract, intellectual or cognoscitive. The same applies to moral debates in terms of benefits and harms, which appeal to utility in the same way. The point lies on the fact that in online environments we should prevent *common harm*. In that sense, utility must be thought of as a value that takes the individual agent into account.

Among the diverse kinds of utility—functional, economical, epistemic, and so on (see van den Hoven 2000, p. 142)—, online technologies used in science and data-searching, as well as the inquiry on information, have many advantages in the epistemic case. Goldman mentions some standards in order to assess the correction of social practices that help us get true beliefs and evaluate their epistemic success (capacity, fertility, speed, efficiency, reliability). Those standards make it easier to observe that, say, reliable epistemic practices make sounder intellectual and cognoscitive practices. That is, they "epistemically reinforce" users and researchers, and increase the power of reasoning and decision making by means of computer devices and communication networks.

However, there are also certain moral rules that can express those values identified by Nagel. It is useful to distinguish two types of rules governing online behavior: the first order and second order moral rules—the latter ones are rules of recognition, which permit us to identify what is moral and what is not.[16] A recognition rule is a rule that serves to recognize properties that we consider acceptable in an action or decision making.[17] There are a lot of *respect rules* in Internet that can belong to conduct codes. Shea (1999) lists ten of them, among which I highlight some of the most relevant for epistemology: (i) learn where you are in cyberspace, (ii) respect the time and the bandwidth of other users, (iii) share expert knowledge and, finally, (iv) do not abuse your abilities and authority.

In professional ethics, to frame permissions and moral duties in terms of responsibilities seems to have more advantages than the abstract moral discourse. This is due to the fact that such permissions and duties can be integrated in the description of social and professional roles, which provides the option for overcoming the gap between cognitive recognition of the truth of some moral propositions and the motivation to act according to them.

---

[16] *"Netiquette"* rules, *rules of respect* and *conduct codes* (say, "don't plagiarize" or "be polite when writing e-mails") are typically second order moral rules.

[17] For instance, "maximize the happiness of the greatest number of people" or "follow instructions."

There have also been some criticisms, usually oriented to the notion of *sin*, against the traditional views about responsibility (Ladd 1988). Given the contextual nature of responsibility—which almost never is a matter of "all or nothing"—, critics think that a correct view would have to make possible the distribution of responsibilities. Hence, a responsibility so conceived would be both relative to the agency and consequentialist. Unlike duties, which take people-specific actions as objects, responsibilities would be understood of as a result of some orientation. In this sense, Schmidtz and Goodin's (1998) proposal must be emphasized with respect to future-oriented responsibility. In economics it is well known that to make a decision brings on a negative externalization when who makes the decision may be a passive subject of the consequences, costs or results of such a decision. The externalization of responsibility occurs when the agents realize that they are responsible for the problems they cause, but at the same time think that the solution to those problems depends on other agents. Instead, responsibility is internalized when the agents take it into account for future consequences of their actions. Online environments, relatively anonymous, appear to induce the "externalizer" attitude in individuals, especially in epistemic responsibility cases. This last point is basic within online practices. We have already seen that information, knowledge, and justification are crucial concepts in people's life. Hence, should we think of the *use* of Internet files (not published yet or, say, with copyright) or of online data-bases as a responsible mode of generating beliefs? Would be it a quite responsible use as to imply a complete moral responsibility if something "would not work" provided an agent acted in virtue of the information acquired that way?

5.4   Technology, Information, and Ethical Codes

New technologies have created some conditions for knowledge and action that imply emergent ethical problems forcing applied ethics to face new moral dilemmas. In online practices, the notion of *information* is a basic one. Cognitive agents depend upon information that, in turn, is based on the development of computer systems. This is the motive for users to be forced to ensure information integrity and to contribute to the common well-being. Hence, here we are within the realm of *information ethics*.

If we can articulate this kind of ethics in terms of an analysis of practices without losing sight of the more abstract reflection, then we have at least two starting points: first, practices will address norms and values that individuals will have to respect and, second, reflection on practices will allow us to elaborate norms and values that will shape the online activity. And among the new problems that an ethics of online knowledge will face, one of vital importance refers to the legal and normative gap existing in the field of certain actions that can be made with computers and Internet, but that previously did not exist—this would be a problem that only online practices determine. As we do not have any previous model with which to confront a case of this kind, we are forced to formulate new moral principles, to develop a new normative, and to find out new ways of thinking about new issues. By and large, in such cases we tend to demand from professionals responsibility in their jobs, to demand that they be ready to solve non-technical ethical questions, and to try to decrease the probability of new ethical problems.

Given the globalized nature of information and communication technologies, it is important to ground professional responsibilities upon both a hard core of values and the respect to all other values. I would wish to mention that there has already been developed an ethical framework for computer science professionals that comprises a list of eight ethical principles and a method by means of which to apply them to particular cases. Simon Rogerson (2001, pp. 308 and f) presents them under the name of "honor," "honesty," "bias," "professional adequacy," "due care," "justice," "social cost considerations," and "effective and efficient action." These are not mutually exclusive principles, of course, but rather constitute an adequate list of ethical aspects that ought to be applied to online practices related to computer systems and uses. That is to say, they may be used to analyze, to report, and to shape practices in all the realms of computer sciences, since their utility for identifying high ethical sensibility areas (*ethical hot spots*) is very relevant.

Almost all those principles are embodied in ethical codes and codes of conduct in professional organizations such as the British Computer Society, the Australian Computer Society, the ACM (Association for Computer Machinery), and the highly influencing Software Engineering Code of Ethics and Professional Practice.[18] However, the most direct underlying question has to do with their *utility*: Are ethical codes really useful in online practices? The comparative analysis of codes can be a good starting point to try to answer this question. Beyond the fields of online information and knowledge, other professional associations (scientists, lawyers, engineers) have also formulated their own ethical codes that can help us identify those online conduct patterns that are advisable and those that are not (see Bengoetxea and Mitcham 2010, Part I). In this regard, applied ethics—applied to information and online knowledge—should depend on the discipline users belong to. This kind of disciplinary fragmentation, furthermore, should not surprise anyone, since it does not appear to be plausible that a *general* epistemic-ethical proposal could comprise all knowledge fields completely. Such a general code would not be very efficient. Therefore, disciplinary fragmentation would fit Nagel's value fragmentation, although it would be set up on at least a minimum range of epistemic-ethical principles common to all professional codes.

Thus, it is important not to divorce epistemology and ethics in the analysis of online practices. If we do so, the risk of systematic plagiarism, false identity, or lack of epistemic authority could lead to the loss of many responsibility requirements. That is why it is necessary to make indicative codes that assist us to take decisions and demand responsibilities from users of online information and knowledge networks.

---

[18] There exist specific information on ethical codes in several texts, among which I'd highlight Berleur and D'udeken-Gevers's (2001) article and the "Appendix V" in the 4th Volume of the *Encyclopedia of Science, Technology, and Ethics* (Mitcham 2005), which provides a series of ethical codes organized by professions and countries.

Acknowledgment This work was supported by the Spanish Government's State Secretary of Research, Development and Innovation (research project: *La explicación basada en mecanismos en la evaluación de riesgos*) [FFI2010-20227/FISO] and partially by European Commission FEDER funds.

References

Bengoetxea, J.B., and C. Mitcham. 2010. *Ética e Ingeniería*. Valladolid: Universidad de Valladolid.

Berleur, J., and M. D'udeken-Gevers. 2001. Codes of ethics: Conduct for computer societies: The experience of IFIP. In *Technology and ethics: A European quest for responsible engineering*, ed. P. Goujon and B.H. Dubreil, 327–350. Lovaina: Peeters.

Brandom, R.B. 1994. *Making it explicit: Reasoning, representing and discursive commitment*. Cambridge, MA: Harvard University Press.

Chalmers, D.J. 2002. Consciousness and its place in nature. In *Philosophy of mind*, ed. D.J. Chalmers, 247–272. Oxford: Oxford University Press.

Clark, A. 2003. *Natural-born cyborgs: Minds, technologies, and the future of human intelligence*. Oxford: Oxford University Press.

Dreyfus, H. 2001. *On the internet*. New York: Routledge.

Gettier, E.L. 1963. Is justified true belief knowledge?". *Analysis* 23: 121–123.

Goldman, A.I. 1967. A causal theory of knowing. *The Journal of Philosophy* 64: 357–372.

Goldman, A.I. 1992. *Liaisons: Philosophy meets the cognitive and social sciences*. Cambridge, MA: The MIT Press.

Hegel, G. W. F. 1988/1821. *Grundlinien der Philosophie des Rechts*. Berlin: In der Nicholaischen Buchhandlung. Spanish version by Juan L. Vermal: *Principios de la Filosofía del Derecho o Derecho Natural y Ciencia Política*. Barcelona: Edhasa

Ihde, D. 1990. *Technology and the lifeworld: From garden to earth*. Bloomington: Indiana University Press.

Johnson, D.G. 1994. *Computer ethics*. Englewood Cliffs: Prentice Hall.

Ladd, J. 1988. Computers and moral responsibility: A framework for an ethical analysis. In *Information web: Ethical and social implications of computer networking*, ed. C. Gould, 207–227. Boulder: Westview Press.

Langsam, H. 2008. Rationality, justification, and the internalism/externalism debate. *Erkenntnis* 68: 79–101.

Mitcham, C. (ed.). 2005. *Encyclopedia of science, technology, and ethics*. Detroit: Macmillan.

Nagel, Th. 1979. The fragmentation of value. In *Mortal questions*, ed. Th. Nagel, 128–141. Cambridge, MA: Cambridge University Press.

Priest, S. (ed.). 1987. *Hegel's critique of Kant*. Oxford: Clarendon Press.

Rogerson, S. 2001. A practical perspective of information ethics. In *Technology and ethics: A European quest for responsible engineering*, ed. P. Goujon and B.H. Dubreil, 305–325. Lovaina: Peeters.

Schmidtz, D., and R.E. Goodin. 1998. *Social welfare and individual responsibility*. Cambridge: Cambridge University Press.

Shea, V. 1999. Netiquettes. http://www.albion.com/netiquette/corerules.html. Accessed on 28 Mar 2013.

Steinhart, E. 1999. Emergent values for automatons: Ethical problems of life in the generalized internet. *Journal of Ethics and Information Technology* 1–2: 1–6.

Thagard, P. 1997. http://cogsci.uwaterloo.ca/Articles/Pages/Epistemology.html#anchor04. Accessed on 28 Mar 2013.

Tunick, M. 1998. *Practices and principles: Approaches to ethical and legal judgment*. Princeton: Princeton University Press.

van den Hoven, J. 2000. The internet and varieties of moral wrongdoing. In *Internet ethics*, ed. D. Langford, 127–157. New York: St. Martin's Press.

Weckert, J. 2000. What is a new or unique about internet activities? In *Internet ethics*, ed. D. Langford, 47–64. New York: St. Martin's Press.

Zagzebski, L. 1999. What is knowledge? In *The Blackwell guide to epistemology*, ed. J. Greco and E. Sosa, 92–116. Oxford: Blackwell.

Chapter 6
Risk, Uncertainty, and the Dimensions of Technological Rationality

Amparo Gómez

6.1 Context

Philosophy of technology has highlighted the practical elements involved in technology such as the rationality of decision-making, choices and actions, and their relevance to understand it. Technology belongs to the sphere of praxis (as well as to the sphere of knowledge); therefore it is closely related with the rationality of actions, decisions and choices that take place within it.

This paper aims to analyse technological rationality in order to develop three aspects: (a) risk and uncertainty are consubstantial to technological rationality and constitute a clear limitation to the idea that technology is a field maximally rational; (b) technology, through its consequences, enters fully into the realm of public and social life, and hence in the field of politics; therefore, the technological rationality is not merely reduced to a question of efficacy or effectiveness; and (c) its consequences place technology in the sphere of responsibility, thus, technological rationality is not just a question of means, it is also a question of ends.

6.2 Instrumental Rationality and Criteria of Effectiveness and Efficiency

Rationality is a key component of technology since it is a necessary condition for its efficacy and effectiveness. Although different proposals on technological rationality have been developed, it has been generally understood that, mainly and primarily,

A. Gómez (✉)
Faculty of Philosophy, Campus of Guajara, University of La Laguna,
La Laguna (Tenerife) 38296, Spain
e-mail: agomez@ull.es

© Springer International Publishing Switzerland 2015   105
W.J. Gonzalez (ed.), *New Perspectives on Technology, Values,
and Ethics*, Boston Studies in the Philosophy and History of Science 315,
DOI 10.1007/978-3-319-21870-0_6

technological rationality is concerned with the efficient and effective use of means to achieve certain ends. Efficacy is relative to the achievement of goals sought (it refers to the capacity to achieve an objective), while effectiveness relates to the choice of the best mean to achieve the goal sought, hence, with the best fit between the objectives and the results of an action.[1] In Ellul words, technology should be taken to mean "the totality of methods rationally arrived at and having absolute efficiency" (Ellul 1964, p. xxvi). Therefore, technological rationality is concerned with determining the optimal means to achieve efficiently a pre-established end.[2] Technologists would have all the relevant knowledge regarding the alternatives (future states of the world), their preferences would be rational and they might determine which alternative would be the best to achieve the intended objectives. As has been noted by Herbert Simon, once the goals have been established, the agents merely limit themselves to choose the (optimum) option; the ends pursued are the only elements which may vary, but they are already given to the technologist.[3]

Technological rationality, therefore, is instrumental and it involves the cognitive and practical spheres, but not the evaluative one. It cannot tell us where to go; in the best case it can tell us how to arrive. It does not consider unintended consequences, and as Wenceslao J. Gonzalez reminds us "it appears as neutral as regards the ends" (Gonzalez 1998, p. 102). This rationality is specified in the model of rational choice stemming from economy and it is closely linked to the thesis of neutrality that carefully distinguishes between technology and its uses. It is true, as pointed out by Gonzalez, that technological rationality is mediated by economic rationality, however this does not imply that it must be understood as neoclassical maximizing rationality, but in terms of Simon's procedural and limited rationality; technological rationality "cannot aspire to the maximum in the strict sense but only to the *optimum* on the basis of our capacity."[4]

From this point of view, technological rationality is an internal question, which has little to do with external considerations related to the application of technology and its consequences, and even less with its ends. Consequently, it involves a strict separation between internal factors associated with the means and external factors related to the ends and the consequences of technology. Thus, technological development follows an internal, rational and neutral logic, independent of external considerations

---

[1] According to Quintanilla, his strategy is to measure technical efficiency–on the basis more than economic efficiency, the thermodynamic efficiency–as: "a function of the level of adjustment between results and objectives of an action." Quintanilla (2005), p. 220.

[2] As is well known different theoreticians have characterized technological rationality in terms that go far beyond this conception taking into account external rationality and rationality of ends, see for instance, Gonzalez (1998); Mitcham (2005), p. 788; Rescher (1988); and Agazzi (2004).

[3] This has been pointed out by Herbert Simon characterization of substantive rationality. He considers that this is the case because in conditions of certainty, one of the options always emerges as objectively preferable to the others, and this occurs almost deterministically; Simon (1976). See also Simon (1978).

[4] Gonzalez (1998), p. 113. For an analysis of *economic rationality* in technology, see Gonzalez (1998), pp. 95–115. In this work, he states that "economic rationality may serve as a link between scientific rationality and technological rationality," Gonzalez (1998), p. 97.

and, therefore, it is deterministic since it would be determined uniquely by this logic.[5] This means that there are no harmless or dangerous technologies; there are only technologies which can be used for good or for evil. Technologies themselves are not inherently good or bad; it is its use what falls into those categories. Technology is rational in relation to means, and neutral in relation to ends and the consequences of its use (cf. Gómez 2001).

This conception of technological rationality guarantees that technology offers the best possible responses to the problems posed. Of course, mistakes may be made at some point during the process, but this is something that technology itself can resolve by rectifying in accordance with traditional scientific procedure. Therefore, any external consideration regarding the solutions provided must either submit itself to technological judgement or risk being incompatible with the criteria of technological rationality. This leaves social, political or moral judgement at a disadvantage, since their inclusion in the realm of technology would not be rational.[6] The analysis of technology must distinguish between the internal logic of technology, which is neutral and rational, and considerations related to external questions.

6.3 The Consequences of Technology: Intended and Unintended

The influential concept of technological rationality outlined above totally excludes the question about the consequences of technology, mainly regarding the unintended consequences. Consequences are externalities that are simply not taken into consideration. The rational strategy consists of externalising all those variables that are not related to internal questions of efficacy and effectiveness; this enables optimize decisions and actions. The idea is that by proceeding in this way we will achieve our technological objectives, which in turn will bring us closer the best of all possible worlds, in which an adequate technological solution will be provided for each and every problem, including those generated by technology itself.

However, rational procedure in technology has to consider all the variables in play including those related to the dangerous effects of technology and the risks and uncertainty they generate. Acting as if none of these existed is hardly rational, and indeed serves only to increase the risk and uncertainty inherent in technological development, which thus appears blind to its effects.[7]

---

[5] To technological determinism see Gonzalez (2005), p. 30. As argued Gonzalez (2005), pp. 31–32, Niiniluoto holds an interesting middle position between determinism and voluntarism. Also, see Niiniluoto (1994, 1997).

[6] For moral judgement of science and technology, see Agazzi (2004), pp. 127–139.

[7] Nuclear waste continues to be hazardous for a long time; the plutonium in nuclear power plants will always enable the construction of nuclear weapons; genetic defects and alterations may last until the end of time; the heating of the atmosphere seems to be irreversible, etc. For an analysis of scientific progress and technological innovation, see Gonzalez (1997).

Risk and uncertainty are consubstantial to technological development and are related to factors such as:

1. The importance of the time horizon, or in other words, the difficulty of knowing the mid-term and, above all, the long-term consequences of technology. In order to proceed in a rational way it would be necessary to anticipate both the present and future consequences of a new technology (or of the modification of an existing one).

2. The collateral effects of actions which are by-products of intentional actions and are extremely difficult to foresee, since they are simply unknown.[8] The collateral effects which should be anticipated in technology are, above all, the negative ones: an option which seems technologically optimum may have disastrous collateral consequences which the technology was unable to foresee and which may affect the environment, health or life itself.

3. The irreversible nature of the effects produced (both intentionally and unintentionally): once certain consequences have been set in motion, there is no going back.

Risk and uncertainty are a clear limit to the idea that technology is a maximally rational field and that technological development is based on the complete rational control of reality. There are things that escape this control, quite simply because they are either only partially known or not known at all.

6.4 Risk Estimation

Commonly, risk is defined in terms of probabilities, and its estimation is based on statistical methods. The basic idea of the theory of rationality is that risk enables rational decisions on the basis of objective probabilities and the satisfaction of certain axioms. Risk estimation is a question of statistical methods, the application of which provides data which are essential for rational decisions under risk. In fact, as our knowledge of the implications of certain technologies has increased, we have been forced to refine the methods used for calculating risk, developing more sophisticated models to establish its probability. Risk estimation would generate scientific consensus and this consensus would extend to other levels also, including risk management. This would be essential to the assessment and management of risk and making decisions regarding how to proceed about it.[9]

Nevertheless, this standard conception of risk decision and risk estimation poses a number of problems, since: (a) rational decisions under risk fail to satisfy the

---

[8] The basic idea is that many events occur unintentionally, they are the by-products of actions. They may be positive (A. Smith's invisible hand) or negative (perverse effects). For more on this subject, see Elster (1988). Also see, Gómez (2002).

[9] See the interesting proposal for public risk assessment and its legal regulation by Jasanoff (2001). See also Agazzi (2004), pp. 204–205.

axioms of the theory of utility, as has been shown by Simon, Tversky and Kahneman, among others (cf. Simon 1978; Tversky and Kahneman 1973, 1974, 1981); and (b) risk estimation is not a starting point capable of offering incontrovertible data which guarantee scientific agreement (and agreement in all the other spheres: social, political and legal); on the contrary, the situation is most commonly one of disagreement and even controversy.

Focusing on estimation we find that statistical risk estimation lacks methods with universally accepted criteria which provide data that are not indeterminate and, therefore, not open to more than one interpretation. Risk estimation includes judgments and choices which depend on the values prioritized. The result of these estimations depends not only on the methods, judgements and decisions involved in the process (for example, the decision regarding which methods to use), but also on the internal and external values involved in the choices and judgements, as well as in the interpretation of data. As Mayo points out, estimations include choices for which there are no unequivocal scientific answers, and these choices have important political implications and involve specific political positions.[10] Choices and judgments are made in risk estimation itself, and therefore determine to a large extent what is and is not considered as risk.

There are some examples of the types of choices involved in risk estimation, according to Mayo (1997, p. 227):

(a) Epidemiological data:

– What weight should be attached to studies with different results? Should this weight be in keeping with their statistical power?

– What weight should be attached to different types of studies (prospective versus case-control studies)?

– What level of statistical significance should be required for the results to be considered positive (in relation to risk)?

(b) Bio-test data with animals:

– What degree of confirmation of positive result should be required?

– Should negative results be ignored or considered less important?

– Should a study be assessed in accordance with its statistical power?

– How should the occurrence of rare tumours be treated?

– What models should be used to extrapolate the results to humans?

There are some examples of the type of choices involved in the assessment of dose response:

(a) Epidemiological data:

– What dose response models should be used to extrapolate the results from the observed doses to the relevant doses?

---

[10] Mayo (1997), p. 227. Given that there is more than one scientifically acceptable response to these questions, there is more than one plausible choice.

(b) Data from animal tests:

– What mathematical models should be used to extrapolate the results from experimental doses to exposure in humans?

– Should the dose response relationships be extrapolated in accordance with the best estimations or in accordance with the upper confidence limits? If this latter option is chosen, which confidence limit should be used?

The responses given to these questions – and, therefore, the choices made – will determine different results of the risk estimation. These choices reflect different methodological and technical assumptions, different perceptions of the risk and its scope, and different values regarding how to act regarding it. The result is scientific disagreements and controversies regarding the results of risk estimations.

The study of the carcinogenic risks of formaldehyde in humans is an interesting example of the kind of choices and values involved in risk estimation. The interest on the relationship between formaldehyde and cancer in humans arises from the great use of formaldehyde by engineers since it is one of basic organic compounds most used by the chemical industry due to its great power antiseptic and preservative. The influence of formaldehyde on cancer in humans was studied by the Chemical Industry Institute of Toxicology in the US in 1979. The study was carried out with rats and concluded that the formaldehyde implied risk to humans.[11] The conclusion was sent to the Environmental Protection Agency in 1980.

With the change of administration during the Reagan era, the new director of the Environmental Protection Agency, Anne Gorsuch, carried out a second study with humans (workers exposed to this substance). The new study found no evidence of risk of cancer in humans, and the substance was, therefore, not classified as hazardous. The statistical study with humans commissioned by Gorsuch concluded that the increase in the number of cancer cases among workers exposed to the substance was not statistically significant. It was therefore concluded that formaldehyde posed no risk to human health.

When the study was reviewed some years later, many questions were raised regarding the way in which the data had been interpreted, the technical decisions, the values underlying the conclusions, and the pressure exerted on researchers to find no evidence of risk for humans. The review of the study found that while the data obtained were indeed not statistically significant, they were however relevant regarding the existence of cancer risk for humans. The questions raised then, were as follows: (a) what increase is considered statistically relevant in order to affirm the existence of a cancer risk?, (b) does a statistically non-significant result necessarily imply that there is little or no risk?, and (c) In other words, what was the risk content for those individuals exposed to formaldehyde, despite the statistical non-relevance of the results?[12] The answers to these questions were directly related to the importance attached to the fact that some people might end up developing cancer and how this

---

[11] In experiments with animals have been shown that formaldehyde causes cancer in concentrations greater than 6 ppm in the air breathed.

[12] That is what Beck terms the "the *hazardousness of a risk*" the risk of the risk. See Beck (1992), p. 29.

fact was assessed in the two studies. It was this that determined the non-cautious treatment of risk in the first case, and the cautious treatment of risk in the second.

Furthermore, a member of the team which carried out the second study with humans pointed out regarding the conclusion of the first one: "this means that the data are inconclusive, not that they fail to support an association between exposure to formaldehyde and cancer in humans" (Mayo 1997, p. 240). For a cautious perspective the data supported the association, but not for an opposite point of view. Therefore, the review of the first study questioned that a more cautious and protectionist perspective to the risk was not adopted; in order words, the review questioned the values associated with the decision made in the first study with humans. It also questioned the method employed, since unless it could be established that the risk was three times higher in those individuals who had been exposed to formaldehyde than in those not exposed to the substance, the risk did not matter. In fact the review shown that the method used a broad scale which was unable to detect small increments.

In the conclusion of the review, many scientists claimed that the authors of the study with humans had acted *irresponsibly* and *incompetently*. Science had been placed explicitly at the service of politics, inappropriate methodological decisions had been made, and those involved had failed to proceed in a protectionist manner. However, the new conclusions were not unanimously accepted, since many experts believed that statistical relevance was the best instrument available for estimating risk. Therefore, the new results were also the subject of controversy, arising from the fact that the first estimation was not incorrect, the data provided by the second study were not conclusive and the new study was excessively protectionist. In 2004, the OMS ruled in favor of the relationship between formaldehyde and cancer in humans, but making clear that epidemiological studies have not shown any cause-effect relationship in the cases of cancer studied.[13]

This type of disagreement in the estimation of risk can be found in almost all the great areas of technological development. These disagreements become even more acute when decisions must be made regarding how to manage the risk, or in other words, when deciding which measures and regulations should be established regarding a risk which might be in one of this options[14]: (a) is generated by the high level of technological development attained in heavily industrialised countries, (b) is global, since it is no longer linked to the place in which it is generated, but rather extends across borders, populations and generations, (c) is often irreversible and difficult to perceive (it is present in chemical formulas, biochemical formulas, imperceptible products released into the air, water and our own bodies), and (d) can only be detected by scientists and experts.[15]

---

[13] Only USA, Japan and UK maintain preventive regulations.

[14] According to Bradbury, risk management requires the development of institutional procedures for structuring between two different perspectives on risk, the scientific and the social views. Cf. Bradbury (1989), p. 394.

[15] These points have been developed by Beck (1992, 1998).

6.5   The Social Perception of Risk

Scientific disagreement in the estimation of risk has repercussions for the "social perception of risk," giving rise to an increasing gap between scientific, technical and social perception of risk. Mistakes by overestimates and underestimates, disagreements, avalanches of reports and figures, and contradictory or simply incompetent decisions by supposed experts, all result in a loss of credibility in the scientific treatment of risk and an increasingly sceptical and questioning public opinion.[16] Scientists treat this situation as evidence of the irrationality of public opinion, but the problem cannot be dismissed so easily. Firstly, because there are objective reasons for this lack of confidence, and secondly, because the point of view of those exposed to the risks is legitimate. Risks are also social, not just scientific or technical.

The gap between the scientific treatment and the social perception of risk is related to an idea that has already been mentioned in the example of formaldehyde: the probabilistic estimation of risks is one thing, but the dangers faced by the specific individuals, groups and societies who are exposed to the risk and may suffer its consequences is something quite different. Beck tackles this issue by asking the question about what is the risk of the risk (Beck 1992, pp. 29–31). In other words, the question is about what is the *hazardousness of a risk* for those who are exposed to it. This is not a question that can be resolved merely by calculating probabilities, since what is in play here is precisely the meaning of these probabilities. The meaning of probabilities regarding damage is very different for technicians than for citizens or populations at risk, who perceive the threat posed by the risk data in a very different way. It is a very different matter to estimate the probability of an increase in cancer due to exposure to formaldehyde, than to actually be exposed to a substance which may cause us cancer, even if the probability of this happening is not statistically relevant. It is very different to estimate that the objective probability of a nuclear accident is low, than to accept the risk of suffering the damage caused in the event of its occurrence.

Therefore, the scientific estimation of risk and its social estimation do not necessarily coincide because they imply different perceptions and assessments of risk and different values with regard its meaning. This is an important issue regarding managing and regulating risk, since it begs the question: should such activities take into account only the scientific point of view or should they also include the social perspective? The answer to this question locates risk in the political field, in which there is an interweaving of perspectives, values and interests of different kinds which makes the establishment of a single response impossible to that question.[17] However, what this question highlights is the fact that risks have social and political content, and this content poses key questions for technological rationality: who makes the decision on that something is a risk and what must be done about it? Who

---

[16] Beck (1992); the author has developed these issues extensively.

[17] See the interesting Gonzalez work on the role of values in the configuration of technology: Gonzalez (2008, 2016).

makes the decision on what is safe and what would be safer? Who makes the decision on that something is "harmful," who it may harm and how much?[18] These types of questions affect not just practical rationality but cognitive rationality also, since what is at issue is our knowledge of risk and therefore, what we should believe and accept as a risk.[19]

Technology and its effects have entered both the public sphere and therefore political sphere, in addition to the social one, characterised by the sheer number of different perceptions, points of view and values in relation to technological phenomena.[20] The political content of risks is basically related to two factors: (i) the fact that the definition of the risks and their estimation is no longer purely scientific-technical, and (ii) political control of risk. The first issue is linked to the fact that the definition and estimation of risks is sensitive to political positions, and therefore to different values and interests (as seen in the case of formaldehyde). The second is related to political intervention in the management and regulation of risks once they enter the public sphere. The high political potential of risks has given rise to emergence of hitherto nonexistent and now increasingly important areas of political intervention and action.[21] Risk problems and democracy problems appear interconnected to the extent that society itself is demanding the opening of decision making-processes and its right to participate in them. New unexpected actors emerge in the midst of controversies, meaning that the debates are no longer limited to scientists and technologists. These actors demand the right to participate, not only in the management of risks, but in their definition also, as well as in the debate regarding what directions technological development should follow. The debate about nature, about life and the causes of risks, which was once apolitical, has now become political.[22]

Technological rationality finds it harder and harder to ignore the type of question posed by the political and social dimensions of risk. It can no longer limit itself to the assumption of technological efficacy and optimism. Rationality is no longer merely a question of efficacy and effectiveness, it is now necessary also to tackle the assessment of the risks generated by technological development. Rationality is also evaluative, and evaluative rationality and means-ends (instrumental) rationality are two dimensions of technological rationality.[23] However, these two dimensions of technological rationality do not, always, exist harmoniously, in separate spaces. These dimensions of technological rationality can be opposed (if not downright clashing, such as in the case of the scientific and social perception of risk), giving rise to debates which aim to find rational answers to questions such as: what should

---

[18] For these questions see Balmer (2013).

[19] Rescher distinguishes between the two forms of rationality in relation to science; see Rescher (1988), pp. 2–3.

[20] For the relationship between science, technology and politics see Gómez and Balmer (2013). Also at the cultural sphere, see Mitcham (1980).

[21] According to Beck. See Beck (1992), pp. 36, 48–49, 52 (among others). See also Beck (1998).

[22] This is a key idea in Beck's analysis of risk society; see Beck (1992), passim, and Beck (1998).

[23] According to Rescher, the three spheres of the technological rationality are: The cognitive, the practical and the evaluative ones. See Rescher (1988).

take priority, the efficacy of a technology for the achievement of a specific objective or the negative social and political assessment of the risks that technology poses? It is for this reason that the proposal of rational criteria for making decisions in this type of situation has occupied a key place in recent thinking about technological rationality, especially in situations of uncertainty regarding its effects.

6.6 Rationality and Prudence

Technological rationality is also related to the uncertainty of some of its hazardous consequences. Uncertainty arises when the effects of a technology in the short or long term cannot be estimated with objective probabilities, as they can in the case of risks. It is not that we know nothing (uncertainty does not imply total ignorance), it is just that the knowledge we possess does not allow us to establish objective probabilities regarding consequences. In this case, the rational way of proceeding is limited to the subjective probabilities. However, this is considered an excessively weak form of rationality, which is why rational criteria are introduced. Upon these criteria is based the rationality of the decisions and choices made in conditions of uncertainty.[24] The question, then, is how to rationally base the choice of criteria, to decide and act rationally under uncertainty regarding the dangers generated by the development of certain technologies.

Various authors have responded to this question by proposing prudent criteria for dealing rationally with uncertainty regarding technological hazards. One example of this is the proposal forwarded by J. Elster, who claims that in situations of uncertainty, the rational thing is "acting as if the worst will happen" (1983, pp. 203–204). It must be decided by comparing and assessing the worst consequences that may arise, since however small the possibility of a disaster of infinite dimensions, its effects would be infinite; so one must decide as if this were really going to occur.[25] The criterion proposed by Elster has important implications for technological development, since it acts as a recommendation from which we can derive that proceeding with caution in relation to certain technological developments is the most rational choice.

Diverse authors have highlighted the importance of a cautious approach and of prudence, opposing the technological imperative which holds that "everything that is technologically possible should be done." In response to the magnitude of our technological capacity, Hans Jonas talks about the need for a "heuristic of fear," which consists of always taking the worst consequences into account (Jonas 1979). During the middle of the last century, Jacques Ellul also proposed what he termed the "ethics of nonpower" (Ellul 1980, pp. 204–212), based on the idea that human beings should accept that they do not necessarily have to do everything of which

---

[24] Both are forms of rational choice in contexts of uncertainty.

[25] Elster (1983), p. 203. This author points out that an infinite number multiplied by any positive quantity (no matter how small), is still infinite.

they are capable, as a means of controlling both technological development and its effects. Something which is technologically feasible may be socially or even morally undesirable due to the danger of its possible consequences. The aim, then, is to avoid the negative consequences of actions which despite being possible may nevertheless have harmful effects (cf. Gómez 2007). Hence the importance attached to the regulation of many technological advances, even though this regulation may imply certain limitations.[26]

From World War II onwards, scientists themselves recognised the potentially adverse implications of some of their research (mainly in relation to nuclear power and its military use). This acknowledgement sparked a debate on scientific responsibility and the importance of providing society with scientific training. In 1947 the *Engineers' Council for Professional Development* developed the first ethical code in engineering pledging "interest in the public welfare." In subsequent revisions (1963 and 1974) this commitment to human welfare was reinforced (Mitcham 2016).

Also during the 1970s, the *National Science Foundation* (the public agency responsible for funding research in the US) decided to establish funding specifically for research aimed at reinstating the importance of ethics in science and technology. One consequence of this was the development of the program on Science, Technology and Society.[27] The debate which raged during the 1950s about nuclear weapons tests resulted in the *Partial Nuclear Test Ban Treaty* in 1963. On the other hand, the biologist Rachel Carson demonstrated the destructive capacity of the excessive use of pesticides in her book *Silent Spring* (1962). This led to the establishment in 1970 in United States the *Environmental Protection Agency*, a governmental organisation which was to gain considerable international influence.

After 1970, in response to the environmental consequences and risks of genetically-modified organisms, the scientific community adopted a voluntary world moratorium on this type of technology, with the aim of establishing appropriate protocols for its development. A certain level of agreement was reached between scientists and politicians regarding the rejection of a strict *laissez faire* in technologies whose effects were uncertain. This was reflected in the political sphere (at least in the European Union) by the adoption of the *precautionary principle* and by regulations governing certain technologies.[28] The idea that underpins the precautionary principle is that any new technology should be considered as if it were dangerous until proven safe.

However, the adoption of the precautionary principle in Europe, as noted by Mitcham, was "faced with the overriding *ethos* of enthusiasm for technology still emerging in other parts of the world market; it was difficult to implement in a pluralist society and was separate from any remnant of traditional ways of life that may have truly supported it. Moreover, in specific political debates it is hard to know

---

[26] Prudential recommendations are located within the framework of consequentialist ethics based on the principle of responsibility that is today fairly popular; for more on this subject, see Agazzi (1999).

[27] In the US in the 1980s.

[28] In relation to the precautionary principle, see Foster et al. (2000). See also Gómez (2003).

how safe is safe enough" (Mitcham 2005, p. 788). On the other hand, the application of the precautionary principle has faced also problems of collective action. These types of problems show the difficulties for countries adjusting to prudent criteria or a cautious approach. Why would some countries decide to suspend or limit the development of certain technologies – as for example, biological and chemical technology related to the production of weapons of mass destruction – when they are uncertain about what other countries will do? The consequences of a chemical or biological war or disaster are not substantially altered because a country – or a few countries – suspend the production of these technologies and hence weapons of mass destruction. These countries will suffer the consequences of a disaster just like the others, while are paying the cost of such suspension. To adopt a policy of not developing nuclear energy (as announced France in 1990s) is subject to this dilemma. Even if a country decides unilaterally to suspend nuclear energy production, the risk of a nuclear disaster (or a nuclear warfare) does not decrease substantially. The question is then: why pay the cost of abandoning nuclear energy? The best choice is collective to eliminate nuclear energy due to its consequences, but this will not happen unless all parties reach an agreement and act collectively since the best individual choice is to continue producing this energy.

The solution to this type of situation is the negotiation and the achievement of agreements for cooperation through transnational organizations to stop the research and production of such technologies. But the achievement and fulfillment of such agreements is also not easy, as experience shows. The agreements are difficult to achieve and once attained are subject to the free rider problems that can block them and make them inoperative. Briefly stated, the problem arises when one (or more) of the parties break the agreement (continues to develop dangerous technology) and they benefit from the compliance to agreement by the others, without paying the cost of doing it. This problem is well illustrated by the difficulties faced by international agreements like the Kyoto Protocol, the Cartagena Protocol, or packages of measures Erika I and Erika II.[29]

On the other hand, a cautious approach has not generated a general consensus within the scientific community (neither among the other social actors). Three types of arguments have been raised against the cautionary approach: (a) that it hinders innovation and paralyses research; (b) that this approach overlook the importance of technological advantages and benefits, and their positive effects[30]; and, (c) that the control or limitation of technology may involve more costs than its negative effects.

---

[29] The solutions to these situations are of two types: Strengthening compliance of the international agreements with sanctions and/or rewards–which is not very feasible at certain levels since it would be required strong transnational organizations which carry out this task–or appeal to moral reasons about what should be done independently of what others do–which has obvious limitations.

[30] We forget, for example, that the genetic manipulation of food often renders it cheaper, improves its quality and extends the areas in which it can be grown, which has important implications for the developing world, it also provides a solution to the problem of food security in the world and it renders plants more pest-resistant, thus lessening the need for pesticides; that nuclear power is the

Hence, to act in accordance with protectionist principles may generate social and political resistance. One example of this was the announcement by Germany – at the end of the 1990s – that it planned to renounce nuclear energy and suspend the shipment of its nuclear waste to France and the United Kingdom.[31] The news triggered a political and economic uproar in France and the UK, for whom the suspension was seen as a serious mistake which threatened the nuclear industry, and would have grave economic and labour costs. Indeed, in France, sectors operating within the waste recycling industry would suffer serious economic losses. All this resulted in a political face-off between the governments of these countries, since France demanded compensation from Germany (it was calculated that the French public sector company alone would lose 30 billion francs over the next 10 years). The issue was debated in the French National Assembly, which reiterated its commitment to nuclear power, highlighting the negative consequences of its elimination: job losses, use of other energy sources which increase the concentration of CO_2 in the atmosphere, dependence on countries in the Persian Gulf and Russia, etc. It is important to point out that only 30 % of Germany's energy came from nuclear power, whereas this figure was 80 % for France. Diverse social actors in France opposed the step being taken by Germany, including certain workers' groups: Cohn Bendit–the famous Danny the Red–was threatened with iron bars by workers from The Hague when he went to the city to give a talk in favour of abandoning nuclear energy.

What this example shows is that there is no consensus regarding prudential criteria, and the consequences of applying them may be assessed as negative by diverse social actors. There is therefore no unanimous social perception of these measures – just as there is no unanimous social perception of risks and their regulation –, either politically or scientifically. The complexity of the decision-making regarding which criteria should be followed in situations of uncertainty is evident, and the rationality of this type of decision depends both on the scientific and technical arguments and of the assessments of the various actors implied in the decision-making process.

Finally, the question regarding what should be done with the risk and uncertainty, inherent to the high-level technological development and its dangerous consequences, involves to the ends of technological development.[32] This in turn introduces the question of rationality of ends, within the framework of technological rationality.

---

cheapest and least polluting energy source currently available; and that biotechnologies have important medical applications, etc.

[31] The announcement appeared in the Spanish newspaper *El País* on 24 January 1999.

[32] According to van de Poel, technologies are inherently normative and include instrumental and final values. The technical function of certain technologies is to prevent certain evils and dangers being instrumental regarding moral values. The example is a seawall for preventing flooding and to achieve security for people as part of its function. See van de Poel (2016).

6.7 The Rationality of Ends

The key question regarding the rationality of the ends is how to decide rationally which ends should govern technological development and, therefore, how such ends should be established.

The answer to this question has been based on the belief that decisions regarding ends should be democratic. What is chosen should be the preference of the majority – or the expression of what each and every one prefers or a consensus attained on the basis of decisions which should be made in egalitarian, plural and discursive interaction processes.[33] Underlying this approach is the acknowledgement that, alongside scientific, technological and economic factors, the demands of society must also be taken into consideration when deciding on the ends of technological development. A rational technological development is not possible unless attention is paid to social engagement processes, since the direction taken by that development affects us all. The basic idea is that it is not enough to *understand* science and technology, it is also essential to *take charge* of them, and no one should be excluded from this undertaking. This requires establishing participation mechanisms for the different social actors in the contexts in which relevant decisions are made. It is then necessary to accept the possibility of an innovative political activity capable of involving new actors and opening up new options to strengthen the capacity of social participation in the decision-making process (Gómez and Balmer 2013).

Another key idea in this field is that decisions regarding ends should include considerations about social relevance, satisfaction of social needs, public utility or general interest. In other words, they should be based on generally accepted values that constitute a common minimum of public space, and should be open to dialogue.[34] These values will enable rational and responsible technological development, also from the perspective of the ends related to the common good. Nevertheless, this proposal is not without its difficulties, since ideas regarding the *common good* – and the other values aforementioned – actually present differences, which is why these values may result in different proposals. However, the inclusion of considerations related to these types of values is essential when the question is the ends of technological development.

6.8 Concluding Remarks

The analysis carried out here has aimed to demonstrate that technological rationality is not limited to internal considerations about means-ends in a neutral space unaffected by economic, social, political or moral factors, which would be relegated

---

[33] Kitcher approach is inspired by the social choice theory related to social welfare functions, therefore, the final result, which is chosen, must be an expression of what everyone prefers and choose. For Longino decisions have to be taken in discursive interactive processes, egalitarian and plurals which produce a consensus as result. See, Kitcher (2002a), (b); and Longino (2002a, b).

[34] See the characterization of internal and external values in technology in Gonzalez (2016).

to the consequences of its use as externalities (see Gonzalez 1999a, b, 2005). The consequences generated by technology raise important external considerations for a technological rationality focused solely on efficiency and effectiveness as the only values in technological decisions and evaluations. The integrated nature of techno-logical development and the inevitable interaction between its internal and external dimensions affects technological rationality. As Simon has shown, the internal environment of technology (research, design, artifact, etc.) depends on the external environment in which this technology operates.[35] Efficiency and effectiveness imply the external environment and its conditions to achieve the ends purported by the technology being developed. Therefore, the concept of instrumental rationality becomes too narrow; technology and its development become unintelligible if we leave aside another dimensions of technological rationality including rationality of ends. Neither technological research nor technological production can be totally separated from its uses and its consequences, just as they cannot be separated from the ends pursued.

This does not mean that conceptually one cannot distinguish between efficiency-effectiveness–with its rationality of means–, and the evaluation of consequences and the ends of technology–with its assumption of avoiding the dangers and democ-ratizing decision-making on technological development and its ends. However, this distinction does not imply that efficiency and effectiveness constitutes the only form of technological rationality, nor that technological rationality is not affected by con-siderations related to the rational assessment of its consequences and ends.

Acknowledgments This paper has been written thanks to the support of the Spanish Ministry of Economy and Competitiveness, Research Project FFI2012-33998. I am very grateful to Wenceslao J. Gonzalez for his insightful comments and suggestions on earlier drafts of this paper. Thanks are extended to Brian Balmer for reading an earlier draft of this article.

References

Agazzi, E. 1999. Límites éticos del quehacer científico y tecnológico. In *Ciencia y valores éticos*, ed. W. J. Gonzalez, monographic issue of *Arbor* 162: 241–263.

Agazzi, E. 2004. *Right, wrong and science. The ethical dimensions of the techno-scientific enter-prise*. Amsterdam: Rodopi.

Balmer, B. 2013. *Secrecy and science. A historical sociology of biological and chemical warfare*. Burlington: Ashgate.

Beck, U. 1992. *Risk society: Towards a new modernity*. London: Sage.

Beck, U. 1998. *World risk society*. Cambridge: Polity Press.

Bradbury, J. 1989. The policy implication of differing concepts of risk. *Science, Technology and Human Values* 14(4): 380–399.

Carson, C. 1962. *Silent spring*. Boston: Houghton Mifflin.

Ellul, J. 1964. *The technological society*. New York: Vintage.

---

[35] As Simon has argued from an evolutionary approach to technological rationality. See Simon (1983), p. 72. For this topic, see Neira (2012), p. 345.

Ellul, J. 1980. The ethics of nonpower. In *Ethics in an age of pervasive technology*, ed. M. Kranzberg, 204–212. Boulder: Westview.

Elster, J. 1983. *Explaining technological change. Studies in rationality and social changes*. Cambridge/London/New York: Cambridge University Press.

Elster, J. 1988. *Sour grapes: Studies in the subversion of rationality*. Cambridge/London/New York: Cambridge University Press.

Foster, K.R., P. Vecchia, and M.H. Repacholi. 2000. Risk management: Science and the precautionary principle. *Science* 288: 979–981.

Gómez, A. 2001. Racionalidad, riesgo e incertidumbre en el desarrollo tecnológico. In *Filosofía de la Tecnología*, ed. J.A. López Cerezo, J.L. Luján, and E.M. García Palacios, 169–187. Madrid: OEI.

Gómez, A. 2002. Estimación de riesgo, incertidumbre y valores en Tecnología. In *Tecnología, civilización y barbarie*, ed. J.M. De Cozar, 63–85. Barcelona: Anthropos.

Gómez, A. 2003. El principio de precaución en la gestión internacional del riesgo. *Política y Sociedad* 40(3): 113–130.

Gómez, A. 2007. Racionalidad y responsabilidad en Tecnología. In *Los laberintos de la responsabilidad*, ed. R. Aramayo and M.J. Guerra, 271–290. Madrid: Plaza y Janés.

Gómez, A., and B. Balmer. 2013. Ciencia y política una cuestión de fronteras. In *Estudios políticos de la ciencia. Políticas y desarrollo científico en el siglo XX*, ed. A. Gómez and A. Fco Canales, 15–34. Madrid: Plaza y Valdés.

Gonzalez, W.J. 1997. Progreso científico e innovación tecnológica: La 'Tecnociencia' y el problema de las relaciones entre Filosofía de la Ciencia y Filosofía de la Tecnología. *Arbor* 157(620): 261–283.

Gonzalez, W.J. 1998. Racionalidad científica y racionalidad tecnológica: La mediación de la racionalidad económica. *Agora, papeles de Filosofía* 17(2): 95–115.

Gonzalez, W.J. 1999a. Valores económicos en la configuración de la Tecnología. *Argumentos de Razón Técnica* 2: 69–96.

Gonzalez, W.J. 1999b. Ciencia y valores éticos: De la posibilidad de la Ética de la Ciencia al problema de la valoración ética de la Ciencia Básica. *Arbor* 162(638): 139–171.

Gonzalez, W.J. 2005. The Philosophical Approach to Science, Technology and Society. In *Science, technology and society: A philosophical perspective*, ed. W.J. Gonzalez, 3–49. A Coruña: Netbiblo.

Gonzalez, W.J. 2008. Economic values in the configuration of science. In *Epistemology and the social*, Poznan studies in the philosophy of the sciences and the humanities, ed. E.G. Agazzi, J. Echeverría, and A. Gómez, 85–112. Amsterdam: Rodopi.

Gonzalez, W.J. 2016. On the role of values in the configuration of technology: From axiology to ethics. In *Technology, values, and ethics*, Boston studies in the philosophy of science, ed. W.J. Gonzalez, 3–27. Dordrecht: Springer.

Jasanoff, Sh. 2001. Ordering life: Law and the normalization of biotechnology. *Politeia* 17(62): 34–50.

Jonas, H. 1979. *The imperative of responsibility: In search of an ethics for the technological age*. Chicago: Chicago University Press.

Kitcher, Ph. 2002a. The third way: Reflections on Helen Longino's The Fate of Knowledge. *Philosophy of Science* 69(4): 549–559.

Kitcher, Ph. 2002b. Reply to Helen Longino. *Philosophy of Science* 69(4): 569–572.

Longino, H. 2002a. Science and the common good: Thoughts on Philip Kitcher's Science, Truth, and Democracy. *Philosophy of Science* 69(4): 560–568.

Longino, H. 2002b. Reply to Philip Kitcher. *Philosophy of Science* 69(4): 573–577.

Mayo, D. 1997. Sociological versus meta scientific view of technological risk assessment. In *Technology and values*, ed. K. Shrader-Frechette and L. Westra, 217–250. Lanham: Rowman and Littlefield Publishers.

Mitcham, C. 1980. Philosophy of technology. In *A guide to the culture of science, technology and medicine*, ed. P. Durbin, 282–363. New York: The Free Press.

Mitcham, C. 2005. Technology and ethics. In *Encyclopedia of 20th-century technology*, vol. 2, ed. C.A. Hempstead, 785–789. New York: Routledge.

Mitcham, C. 2016. Rationality in technology and in ethics. In *Technology, values, and ethics*, Boston studies in the philosophy of science, ed. W.J. Gonzalez, 63–87. Dordrecht: Springer.

Neira, P. 2012. *Análisis de la racionalidad tecnológica en Internet: De la caracterización filosófica de la Tecnología al estudio de las TICs*. Dissertation, University of A Coruña, mimeo.

Niiniluoto, I. 1994. Nature, man, and technology – Remarks on sustainable development. In *The changing circumpolar north: Opportunities for academic development*, vol. 6, ed. L. Heininen, 73–87. Rovaniemi: Arctic Centre Publications.

Niiniluoto, I. 1997. Límites de la Tecnología. *Arbor* 157(620): 391–410.

Quintanilla, M.A. 2005. *Tecnología un enfoque filosófico y otros ensayos de Filosofía de la Tecnología*. Madrid: FCE.

Rescher, N. 1988. *Rationality. A philosophical inquiry into the nature and the rationale of reason.* Oxford: Clarendon.

Simon, H.A. 1976. From substantive to procedural rationality. In *Method and appraisal in economics*, ed. S.J. Latsis, 129–148. Cambridge/London/New York: Cambridge University Press.

Simon, H.A. 1978. Rationality as process and product of thought. *American Economic Review* 68(2): 1–16.

Simon, H.A. 1983. *Reason and human affairs*. Stanford: Stanford University Press.

Tversky, A., and D. Kahneman. 1973. Availability: A heuristic for judging frequency and probability. *Cognitive Psychology* 5: 207–232.

Tversky, A., and D. Kahneman. 1974. Judgment under uncertainty: Heuristics and biases. *Science* 185: 1124–1130.

Tversky, A., and D. Kahneman. 1981. The framing of decisions and the psychology of choice. *Science* 211: 453–458.

van de Poel, I. 2016. Values in engineering and technology. In *Technology, values, and ethics*, Boston studies in the philosophy of science, ed. W.J. Gonzalez, 29–46. Dordrecht: Springer.

Chapter 7
Biotechnology, Ethics, and Society: The Case of Genetic Manipulation

Vicente Bellver Capella

Biotechnologies are transforming human existence and their potential ever increasing. The discovery of the DNA double helix in 1953 paved the way for genetic medicine and even the possibility to alter the genetic makeup of human beings. Since then, the capacity of the human being to learn about and intervene in their own biological makeup has not ceased to grow: artificial reproduction techniques; cell culture and transplant that allow human tissue and organs to be repaired; developments in nanotechnology applied to healthcare for diagnosing, therapies or rehabilitation; and so on. This power raises both hopes and fears. It can undoubtedly contribute to the well-being of people and human progress, but it can also have undesirable outcomes in the form of known risks, or hidden transformations of human life which have not been decided by anyone.

This chapter attempts to offer an overview of biotechnologies applied to human life, tracing its development since the 1950s to the present day and its close link to society. Accordingly, it is divided into two parts. The first one is concerned with the way in which human biotechnology interacts with society, taking as a starting point some particularly significant events in biotechnology in 2010 and 2011. The second part analyses what could well be the most serious question that biotechnologies pose for human beings: the possibility to completely recreate oneself by these means. Is this the ultimate expression of human emancipation? An incredible dream that can never come true? A plausible option, albeit laden with risks? Or is this something which should not happen under any circumstances?

It is true that the possibility to modify the genetic makeup of a human being has been contemplated and debated since the discovery of DNA, and equally true that this technology is still not available today. But the issue of whether biotechnology

V. Bellver Capella (✉)
Department of Philosophy of Law, Moral and Political,
Campus de los Naranjos, University of Valencia, 46.071 Valencia, Spain
e-mail: vicente.bellver@uv.es

© Springer International Publishing Switzerland 2015   123
W.J. Gonzalez (ed.), *New Perspectives on Technology, Values,
and Ethics*, Boston Studies in the Philosophy and History of Science 315,
DOI 10.1007/978-3-319-21870-0_7

should be sanctioned or not, has once again become the object of attention in recent years with the advent of posthumanist and transhumanist approaches[1] and new debates concerning "human enhancement."[2] However, my aim here is not to deal with all the ethical issues involved, but to simply express some doubts about the solidity of the arguments posed by those who favour the "enhancement" of human beings by means of germline intervention.

7.1 "Biotechnology 2.0"

During 2010 and 2011 there were four events regarding human biotechnology that epitomise some of the social changes that have taken place in this field in the last decade. This section first begins by explaining each in turn. After this I maintain that, in the light of these events, the links between human biotechnology and society break down, into two clearly defined periods to date: the first begins in the 1950s with the arrival of molecular biology[3] and continues until 2000 with the announcement of the decoding of the human genome; and the second, which I propose be called "Biotechnology 2.0," comprises the beginning of the twenty-first century until now.

7.1.1 Genes, "Test-Tube Babies," "Stem Cells," and Clones

The aforementioned four events in biotechnology I would like to address are as follows:

(a) *The tenth anniversary of human genome sequencing.* On June 26 2000, the President of the United States, Bill Clinton, made a White House announcement concerning the first survey of the human genome. He did so accompanied by Francis Collins, Director of the Human Genome Project, and Craig Venter, President of Celera Genomics. Collins led the international consortium of scientists financed with public money, which had been working on the human genome sequence since 1990. Venter, with private funding and using a different sequencing method, had joined the race to decode the human genome 8 years later. The British Prime Minister, Tony Blair, participated in this ceremonious event by satellite link.

Although that was the most talked about announcement, February 2001 saw the simultaneous publication of the human genome map in the two most important scientific journals in the world, *Nature* and *Science*. Hence, the tenth anniversary of this accomplishment was celebrated both in 2010 as well as 2011. It was also later

---

[1] For a review of the philosophical foundations and historical antecedents of post-humanism, see Ballesteros (2007), pp. 21–46.

[2] See Bostrom and Savulescu (2010).

[3] See Sánchez Ron (2000), pp. 253–299.

celebrated in 2013, since the complete map of the human genome was not presented until 2003. These anniversaries were the object of attention of both scientific journals,[4] yet they did not arouse any special interest in public opinion. In 2000 exaggerated and persuasive metaphors were resorted to, and had been used for some years, to underline the magnitude of what had been achieved,[5] pointing out the revolutionary nature of this achievement for medicine. Ten years later, genetics has still not revolutionised medicine and enthusiastic claims have been replaced by questions.

It is acknowledged that the vast amount of information available has proved difficult for hospitals to interpret and that, in short, "Still, genomics and related disciplines are more closely aligned with modern science than with modern medicine" (Harold Varmus 2010, p. 2028). It is also recognised that rather than giant strides to bring genomics to the field of medicine, one has to think in terms of a gradual assimilation of genetic information into clinical practice. What is unanimously accepted is that, as genomics is incorporated into medical practice it will become more personalised, since personal genetic markers–the small differences between the genomes of individuals–will be the decisive markers to fine-tune both diagnostics and treatment.

(b) *The "father" of the world's first "test-tube baby"–Nobel Prize for Medicine.* Another event in 2010 was the Nobel Prize of Medicine awarded to the English physiologist, Robert Edwards. According to the official press release by the Nobel Prize organisers he was awarded "for the development of human in vitro fertilization (IVF) therapy. His achievements have made it possible to treat infertility, a medical condition afflicting a large proportion of humanity including more than 10 % of all couples worldwide".[6]

As is well known, the world's first "test-tube baby," Louise Brown, was born in 1978 as the result of in vitro fertilisation techniques used by doctors Edwards and Steptoe. Since then, there have been around 4 million births across the world using this technique. Some of the techniques which have been developed for more effective, comfortable and safer procedures include: intracytoplasmic sperm injection

---

[4] Both *Nature* and *Science* published editorials and articles by leading scientists to celebrate the anniversary and assess achievements since those dates. The tone used was one of caution, and even reserve, in contrast to the euphoria that surrounded the presentation of the HGP in 2000. See Editorial 2010. This issue included articles by both Francis Collins as well as Craig Venter. In addition, *The New England Journal of Medicine*, the most important medical journal in the world, echoed the anniversary and pondered on the foreseeable development of genetic medicine; Varmus (2010).

[5] Some of a particularly hyperbolic nature, which are now in every-day use are: "the language of life," "the book of life," "the Holy Grail of life," "the language of God," etc. But if we leave aside the religious or theological metaphors, those which reign are of technocratic nature that speak of "programme," "control," "code," "map," etc. For further reading on different types of HGP metaphors and their impact on the citizen health culture, see Davo and Alvarez Dardet (2003).

[6] http://www.nobelprize.org/nobel_prizes/medicine/laureates/2010/press.html (accessed on March 25, 1013).

(ICSI), which replaces the spontaneous penetration of the egg by the sperm; obtaining eggs from the woman's ovary by puncture as opposed to laparoscopy; freezing eggs and ovarian tissue; and preimplantation genetic diagnosis. At present the success rate of these techniques (see de Mouzon et al. 2010) ranges between 20 and 30 %, which some regard as extremely inefficient while the organisers of the Nobel Prize themselves consider this to be a major therapeutic success.

The press release mentioned above highlights the contribution of in vitro fertilisation to solving the problems of so many infertile couples around the world. However, it does not point out the real revolution that followed in its wake: creating human life in the laboratory. Edwards and Steptoe will not go down in history for having put an end to the problem of infertile couples, but rather for having created an alternative reproduction method to that of sexual intercourse between women and men. Infertile couples who undergo assisted reproduction techniques can manage to have a baby, but they are still infertile, and in fact when this infertility is genetic in origin, it is usually passed on to their children. Since 1978, and even though assisted reproduction techniques have helped to bring millions of babies into the world, the problem of infertility has only worsened. Hence, what Edwards and Steptoe achieved was to render human fertility unnecessary in order to have children.

Clearly assisted reproductive technologies (ARTs) have helped circumvent the problem of infertile couples and managed to provide many of them with the child they wanted. But this outcome must be assessed in the general context of some practices that convert procreation into a process which is increasingly subject to human control, and in which many other agents, other than the couple, have become an essential part. Furthermore, this process, as stated earlier, can be used for many ends other than dealing with the problem of infertile couples. For all these reasons, reproduction tends to find itself within the domain of what has been called "wish-fulfilling medicine" (see Buyx 2008; González Quirós and Puerta 2009) and, as a result, a global business (see Spar 2006).

(c) *Geron abandons its clinical trial on, and line of research into, human embryonic stem cells*. In November 2011 the biopharmaceutical company Geron announced its decision to abandon its clinical trial on human embryonic stem cells to combat spinal cord injuries, along with their entire line of research in this field. This did not figure as a news item other than references in the financial sections of the press, however, it was highly symbolic. Geron was, along with Advanced Cell Technology, one of the pioneering companies in research into human embryonic stem cells. Since 1998 it had lobbied for a legal framework in the United States which was more favourable to this research, which had stirred so much debate on ethics there and around the world. Up until the moment Geron announced this decision it was considered to be a strong contender for achieving the first therapies by means of these controversial cells.

A few months after his election in March 2009, President Obama was quick to lift the restrictions governing financing research into human embryonic stem cells.[7] The President understood it was not appropriate to limit scientific research on ideological grounds,[8] and even less so when it concerned a field of knowledge which offered so much promise to put an end to serious illnesses such as Alzheimer's, Parkinson's or diabetes.[9] In the face of this this new scenario, the world pioneering company in embryonic stem cell research wasted no time in requesting authorisation from the Food and Drugs Administration (FDA) to carry out a Phase I clinical trial on people with bone marrow injuries. These experiments received a loan of 25 million dollars from the California Institute for Regenerative Medicine (CIRM), which is mainly sustained by funding from taxes paid by the citizens of California.[10] The trials began in summer 2011 but were abandoned in November of the same year. The reason given by Geron to justify abandoning their line of research into human embryonic stem cells, which had earned them world renown, was based on purely financial grounds. The trial was interrupted along with the entire line of research not because the cellular grafts were dangerous for the trial subjects, but simply because it proved more economically viable for the company to focus its efforts on other lines of research. After this announcement a CIRM press release stated its confidence in the potential of these trials and that perhaps some other company would carry out the project.[11]

(d) *The first triploid embryos are cloned.* In October 2011 the journal *Nature* (Noggle et al. 2011) announced the cloning of triploid human embryos. As opposed to conventional cloning, in which the egg nucleus is substituted for the

---

[7] The two Bush mandates had maintained the prohibition on public financing of research which used cells obtained from human embryos. The most influential scientific journals were highly critical of this measure, on the basis that science was being driven by ideology; see Nisbet, Brossard, and Kroepsch (2003).

[8] "Next, we are restoring science to its rightful place. On March 9th, I signed an executive memorandum with a clear message: under my administration, the days of science taking a back seat to ideology are over. Our progress as a nation – and our values as a nation – are rooted in free and open inquiry. To undermine scientific integrity is to undermine our democracy. It is contrary to our way of life"; http://www.whitehouse.gov/the_press_office/Remarks-by-the-President-at-the-National-Academy-of-Sciences-Annual-Meeting/ (accessed on January 17 2012).

[9] The front cover of *Time* magazine, on February 8, 2009, was dedicated to stem cells with the following heading: "How the Coming Revolution in Stem Cells Could Save Your Life." In issue number 24, January 2009, *Time* echoed approval of the first clinical trial on human embryonic stem cells with the following heading: "Cautious Optimism for the First Stem-Cell Human Trial," http://www.time.com/time/health/article/0,8599,1873825,00.html (accessed on January 17, 2012).

[10] The California Institute for Regenerative Medicine was set up in 2005 to compensate for the lack of public funds set aside for research into human embryonic stem cells by the Bush administration. To date, the only clinical trial with these cells financed by the CIRM was begun and abandoned by Geron. An in-depth and critical monitoring of the work done at this centre since it was set up can be found at http://californiastemcellreport.blogspot.com/ (January 23, 2013).

[11] "Geron discontinues stem cell program, CIRM optimistic about future of stem cell therapies." CIRM, Press release, November 14 2011; http://www.cirm.ca.gov/PressRelease_2011-11-14 (accessed on January 17, 2012).

nucleus of a somatic cell, here the nucleus of the somatic cell was implanted without removing the egg nucleus. The resulting embryo was developed until the blastocyst phase, and the experiment was performed by a team from the New York Stem Cell Foundation Laboratory team, led by doctors Scott Noggle and Dieter Egli.

This event differs significantly from that which figured Woo Suk Hwang. He published two articles in the journal *Science* in 2004 and 2005 in which he stated he had managed to clone human embryos (see Hwang et al. 2004) and had derived specific stem cell lines for patients based on this technique (see Hwang et al. 2005). There are three main differences between these two experiments:

– The first and obviously fundamental difference is that Dr. Hwang's announcement was a fraud, while the results presented in 2011 by Noggle and Egli do not appear to be so.

– The second lies in the fact that, while Dr. Hwang said he had obtained human embryos by means of oocyte enucleation and subsequent transfer of the nucleus of the somatic cell to the oocyte, the new technique performs the transfer of the nucleus of the somatic cell without prior oocyte enucleation.

– The third, and by no means least important, is the contrast between the opaque information given by Dr. Hwang concerning the way the oocytes had been obtained and the total transparency on this point by the team led by Noggle and Egli. In their article they acknowledge that they used oocytes obtained from paid donors (cf. Bellver Capella 2012).

Once again we find ourselves before a situation, as in the previous cases, in which the money factor plays a decisive role in developing research. In order to obtain oocytes donors have to undergo painful treatment and surgery with possible side effects. Many donors are needed to obtain a sufficient number of oocytes to be able to research into nucleus transfer. The most effective incentive to procure donors is to offer a considerable payment by way of compensation for the entailing discomfort and costs for the donor. Whether this practice should be sanctioned or not has been the subject of lengthy debate. In effect, in October 2011 the Nuffield Council on Bioethics published a report on organ donors, which posed the question and argued in favour of compensation for egg donors for research. Along the lines proposed by the British committee on bioethics, the Human Fertility and Embryology Authority, approved a series of guidelines the same October, which established payment of 750 lb for eggs donated in one menstrual cycle.[12]

---

[12] See Human Fertility and Embryology Authority (press release), "HFEA Agrees New Policies to Improve Sperm and Egg Donation Services," http://www.hfea.gov.uk/6700.html (accessed on February 1, 2012).

7.1.2 New Relationships Between Biotechnology and Society

With regard to the relationship between human biotechnology and society, there are some general features can be identified in the light of these four events. Although to a certain degree all these features have been present from the beginnings of contemporary biotechnology, they have taken on specific profiles in the first decade of the twenty-first century.

(a) *The consolidation of the network made up of scientific-technological centres, private enterprise, public administration and public opinion.* Scientists, private companies, public administration and public opinion were ever present in the four events mentioned earlier, and so it is clear that science, and particularly biotechnology, is not a matter which concerns scientists alone. Their research projects and technological developments require increasingly heavier financial backing, and this funding either comes from governments and other non-profit making organisations, or from private companies which invest in these projects for profit. In order to attract financing, researchers have to demonstrate the social interest of their projects in the advancement of knowledge and human progress, or potential profitability for the companies which invest in them.

At least since the 1950s, science ceased to be a pursuit which could be carried out simply by guaranteeing scientists academic freedom in their research. All developments in biotechnology have been the result of a hybrid relationship between science, private enterprise, the State and society, which have been studied in depth by science philosophers and sociologists alike. But in recent decades the pressure to gain access to financing has grown exponentially. The financial resources set aside for R+D have multiplied, but even more so the need for financial backing for science and its technological developments. On the one hand, the infrastructures and human resources necessary to carry out increasingly more complex and ambitious projects generate costs which grow exponentially. On the other hand, the search for zero risk and total safety in these projects causes their costs rocket even more, to the degree that they are difficult to sustain. Research group leaders spend most of their energies raising funds, which the continuity and prestige of their groups depend on. The excellence of the research group is not measured by its scientific results alone, but also by its ability to obtain financial resources. As a result, research and technological development inevitably lean more toward obtaining financing than carrying out research projects based on their own merits. Financing a project can be seen as the best guarantee of its scientific interest. But this is not always the case, regardless of whether we are talking about public or private financing.

Public funding is usually assigned in accordance with a public announcement for proposals, which are decided in line with the evaluations obtained by the projects or research group applicants. However, the leaders of these research groups logically exert pressure so that the reviewers are those who they consider the best suited, so that the lines of research they lead are given a higher priority, and so that the evaluation criteria are the best suited to the projects they present. In the sphere of private financing, including when this is offered through public calls, the main decision-making criteria is usually the short-term profitability of the results of the various research projects or technological developments.

As a result, scientific practice nowadays is determined by the power the leading scientists have over establishing the ground rules (above all when financed from public funds) and by the profitability of the results of the research or technological developments (above all when financed by private enterprise). Consequently the idea of scientific excellence ends up being inverted: it is not what is considered excellent which is financed but rather what receives finance is what is considered excellent.

In this competitive atmosphere to obtain financing one cannot overlook the decisive role played by the media. The media is the main source of information about science for the public and those best placed to guide public preferences when it comes to giving financial support to one or another area of research. The case of research into human embryonic stem cells proves paradigmatic. These cells have still not produced any therapeutic benefits, while adult stem cells have done so since the end of the twentieth century until now. Yet, public opinion is unchanged in that it is cells originating from embryos that are going to be able to regenerate all the damaged tissues in our body. Society's view concerning embryonic stem cells –equally encouraged by scientists, private enterprise and the media– has been decisive in tilting the ethical debate on research into these cells in its favour, passing laws which allow this and assigning huge amounts of money to carry them out (see Nielsen 2008).

(b) *Biotechnologies maximise their "sales" strategies.* I have just pointed out that the economic viability and public opinion backing have become decisive criteria for considering research projects as excellent and, as a result, eligible for financing. Faced with these ground rules biotechnology, like all the science-technology areas in general, has to use the right strategies. One of these, elementary but extremely effective, consists of generating major financial and social expectations. Along these lines, the possibilities of applying certain research results are stressed to the utmost. In contrast, the failure of research results are obscured or massaged, alternatives that could prove more effective are ignored, and the risks and adverse effects are minimised. The research group thus struggles between two forces that are difficult to reconcile: the essential capacity for self-criticism in the advancement of knowledge; and the pressing need to "sell" what you are working on to ensure financing.

This is the case in the four instances mentioned earlier. The Human Genome Project was presented as the definitive step forward towards predictive and personalised medicine. In vitro fertilisation opened the way because it was a solution to a health problem that had become more widespread in recent years and allowed millions of infertile couples to have children. Human embryonic stem cells were presented as the great promise for regenerative medicine and cloning embryos as the ideal technique to deal with the problem of rejection.

However, in the first decade of the twenty-first century it has been proved that these expectations were very often overstated and in some cases fraudulent. When the first draft of the human genome was announced in 2000, Bill Clinton spoke of the beginning of a new era in genetic medicine. Knowledge about the human genome has certainly provided valuable information, but 10 years later it has been acknowledged that genetic medicine is far from being a reality (see Marshall 2011). How many years will be needed to transfer genetic knowledge from the laboratory to the clinic?

There have been noteworthy successes in the case of in vitro fertilisation and, in general, assisted reproduction techniques. However, the impact of these techniques on health and for society has not always been given due attention. There has not been sufficient assessment concerning the degree to which developing these techniques has slowed down research into both preventing and combatting infertility. Neither do we know with any degree of certainty the extent of health problems suffered by children born using ARTs that are associated with these techniques. Lastly, and perhaps most important of all, ART official authorities (specialist journals, scientific bodies, etc.) have lobbied for more flexible regulations, yet in contrast there has been no assessment of the harm done to both women and children (see Annas 2011).

The case of human embryonic stem cells is the quintessential paradigm of misleading expectations.[13] Almost 15 years after they were isolated in the laboratory, and the leading scientific journals in the world baptised this as a scientific landmark opening doors to regenerative medicine,[14] it has not led to a single therapeutic result. There are hardly any clinical trials on this type of cells.[15] The announcement by Geron in 2010, which appeared on the front cover of *Time*, no less, was abandoned a few months after it began. This case highlights an important difference with regard

---

[13] In November 2007 the Shinya Yamanaka team announced they had obtained human induced pluripotent cells (iPS), cells which have the same potential as embryonic stem cells but which were obtained without having to destroy embryos. It is worth noting that the editorial in the *The New York Times* basically consisted of claiming that embryonic stem cells, were like "the gold standard for measuring how valuable the new cells will be." Editorial, "Behind the Stem Cell Breakthrough," *The New York Times*, December 1 2007; http://www.nytimes.com/2007/12/01/opinion/01sat1.html (accessed on January 23, 2012).

[14] See Vogel (1999). In the article he says: "We salute this work, which raises hopes of dazzling medical applications and also forces scientists to reconsider fundamental ideas about how cells grow up, as 1999s Breakthrough of the Year," p. 2238.

[15] Advanced Cell Technology (ACT) was, along with Geron, one of the pioneering companies in working on embryonic stem cells. In January 2012, 2 months after Geron announced it was abandoning clinical trials using these cells, *The Lancet* published a study on the first positive results of a clinical trial with human embryonic stem cells financed by ACT to treat certain eye injuries. The experiment was carried out by two people. There is a certain degree of doubt surrounding the trial since the sponsor, "has been criticized in the past for overstating results, in part because it has been desperate to raise money to stay in business"; Andrew Pollack (2012) http://mobile.nytimes.com/2012/01/24/business/stem-cell-study-may-show-advance.html (accessed on January 23, 2013).

The clinical trial is presented as "the first description of hESC-derived cells transplanted into human patients"; Schwartz et al. (2012), In: http://download.thelancet.com/flatcontentassets/pdfs/S0140673612600282.pdf (accessed on January 28 2012).

to the Human Genome Project. Although the HGP was presented as road to a new way of practising medicine, it never promised short term cures. In contrast, the proponents of embryonic stem cells were quick to offer a convincing discourse laden with therapeutic promises, which allowed them to gain wide support from public opinion and in particular from associations for serious pathologies such as diabetes, Alzheimer's or Parkinson's, in the face of the ethical objections posed by the use of human embryos.[16]

Finally, there has also been a profound disappointment as regards cloning embryos. After the announcement in 1997 that the sheep Dolly had been cloned it seemed that cloning human embryos was just a matter of time. While efforts focused on the birth of a cloned human generated almost universal apprehension, cloning to provide human embryos to use in research into stem cells received wide support throughout the science community,[17] while public opinion in general was more divided.

At present we find that there are three types of deceptions in the field of cloning. The first is the case of the sadly famous doctors Antinori and Zavos, or the Raelian sect.[18] They announced that they had managed to clone embryos and the first human clone was about to be born. Although they generated certain notoriety at that time around the world, their deception is no more than a naïve attempt since they were never able to prove anything scientifically. The second and certainly the most serious, was the case of Dr. Hwang mentioned earlier, who was able to pull the wool over eyes of the journal *Science* and, with this, the scientific community and world public opinion. The third consisted of presenting human embryonic cloning for scientific ends as something that was ethically innocuous, and radically different from cloning for producing humans.[19]

The only significant progress made in this field until now is that mentioned earlier in 2011. As opposed to the earlier research teams, the authors of this experiment went to great lengths to avoid false expectations and were quick to state that the end purpose behind creating triploid embryos was exclusively for research purposes, not therapeutic. Additionally, they openly acknowledged that the eggs used to clone

---

[16] For an interesting review of the background and main arguments that have dominated debate on human embryonic stem cells since 1998, see Nielsen (2008).

[17] The editor of *Science* published a passionate report urging the House of Congress not to legislate against cloning humans as, "it would interdict a wide range of experimental procedures that might, in the near future, become both medically useful and morally acceptable"; Kennedy (2001), p. 745.

[18] Although almost forgotten now, when they are remembered, they seem more like vendors at a trade fair than premier league scientists. We must not forget that these people were called by the National Academy of Science in the US to speak at a symposium and their statements were given prime space major newspapers around the world such as the *New York Times*; see Stolberg (2001), http://www.nytimes.com/learning/teachers/featured_articles/20010809thursday.html (accessed on January 18, 2012).

[19] From the pages of *Science* came the call to avoid the term cloning when referring to the nuclear transfer aimed at obtaining embryos for research into stem cells, and only use this term to refer to cloning embryos to be used for giving birth to human clones; see Vogelstein et al. (2002).

embryos came from women donors who were paid.[20] Here we have a style of scientific statement which is radically different from that followed until then in this field, and, in general, in biotechnologies: less pretentious and more transparent.

(c) *Globalisation minimises the role of Law concerning biotechnologies.* One of the social changes in the last 15 years which is having the most impact on biotechnologies concerns the role played by Law in this field. Up until the 1990s Law exercised regulatory control over biotechnologies at two levels. At a national level, each State fixed the limits within which biotechnology could be developed. Hence, between the end of the 1980s and the beginning of the 1990s, many States promoted citizen debates and parliamentary enquiries which later led to passing laws regulating ARTs. This generated a wide range of positions, from the most restrictive to the most permissive, while some countries were either unable or unwilling to pass any laws on this matter. A similar situation came about in the field of genomic research, with the difference that the very same Human Genome Project (HGP), led by the United States, sat up a permanent working group on the ethical, legal and social implications of this research (Ethical, Legal and Social Implications, ELSI working group).

At an international level, it was understood that States could agree on legal regulations that compiled the fundamental principles that should regulate biotechnologies applied to human life and that, in any event, would safeguard the rights of those people affected: human dignity, right to life, right to privacy, right to informed consent, freedom of research, right to the environment, etc. At a world level UNESCO took the initiative in the field of genetics and in 1997 passed the Universal Declaration on the Human Genome and Human Rights. At a regional level, the role played by the Council of Europe is particularly noteworthy via its Steering Committee of Bioethics. The main outcome was sanctioning the European Convention on Human Rights and Biomedicine in 1997 and, to date, four additional protocols: on the Prohibition of Cloning Human Beings (1998), on Transplantation of Organs and Tissues of Human Origin (2002), on Biomedical Research (2005), and on Genetic Testing for Health Purposes (2008).[21] This is the most comprehensive body of international regulations in the area of biotechnology and the most binding for those States which ratify them.[22]

---

[20] It should be remembered that in the false cloning announced by Dr. Hwang the information about the way in which the eggs used in his experiments had been obtained was at first conspicuous by its absence, and was only obtained after the investigation undertaken by the Seoul National University, where he worked.

[21] This last one has not yet entered into force.

[22] Regulations with effect at a supranational level to guarantee human dignity and human rights with regard to biomedicine, which is the purpose of the Convention and its additional protocols, have not been generally approved. Many sectors have criticised its efforts: those which put State sovereignty before international regulations; those who consider that in the area of science and technology the legal systems should give primacy to scientific self-regulation; those who consider that the term human dignity is of no use and it makes no sense to set up international regulations to protect it; etc. See Mori and Neri (2001).

But this scenario of relative stability in developing national and international regulations on biotechnology applied to human beings, went up in smoke after the end of the 1990s. Since then on the situation has remained as follows:

– As regards laws at the state level, we are witnessing an accelerated loss of effectiveness to regulate practices in biotechnology influenced by two factors: first, the free movement of people means that scientists can work wherever there are laws and financing suited to their research, and, second, individuals can go to wherever certain services are offered which are prohibited in their own country. One such example in the former case is the United Kingdom, which attracted financing and researchers from other European countries where laws governing research into human embryonic stem cells were more restrictive. An example of the latter case is Spain, which has extremely permissive laws regarding ARTs, and has become a reproduction tourist resort for couples or single women wanting to have children from countries where there are legal restrictions (cf. Pennings 2004, p. 2690; Inhorn and Patrizio 2009).

– International Law is equally weakened. There are two alternatives for States to reach agreement on biotechnology. The first consists of passing a body of principles general enough so that any State could feel comfortable with them. In these cases the countries with the most permissive laws manage to obtain a legal framework at an international level which they have already sanctioned for themselves. The second aspires more in the direction of regulating matters and in such a way as to effectively guarantee human rights and dignity. In this case we find ourselves before wide range of State averse to following this regulation. Somewhere between these two options, both the Council of Europe and UNESCO are trying to walk a third path, which is not limited to ratifying that which is prohibited by the most permissive of the States as a minimum universal ethical threshold, but that neither establishes thresholds that are only going to be backed by a few States. The most recent result of such attempts was the Universal Declaration on Bioethics and Human Rights, which has been criticised from those who consider it to be completely inadequate as well and also from those who consider it to be excessively restrictive or lacking in legal basis (see Levitt and Zwart 2009; Schuklenk 2010).

7.2 A "Brave New World," Enhanced Individuals and a Posthuman Future

Although the organisers of the Nobel Prizes insisted that the Nobel Prize for Medicine was awarded to Dr. Robert Edwards because of his contribution to the problems of millions of infertile couples (patent function), one cannot overlook the far-reaching effects that ARTs applied to human life have had on the power of humans over the future of the human race.

ARTs have made a decisive contribution to considering human embryos as an object of experimentation, and furthermore put thousands of frozen embryos in the hands of researchers that were never going to be implanted in a woman. The concurrence of these two factors facilitated the development of research into isolating human embryonic stem cells in the laboratory.

But ARTs were not only responsible for generating innumerable "spare" embryos and helping to sanction the idea that human embryos could be used for research under certain conditions. They made it equally clear that human reproduction was something which could take place outside intercourse. It could be a process that was subject to quality control in a laboratory. And, in this case, the characteristics of future children no longer needed to be left to random genetics and instead could be chosen by the progenitors. The way this choice is made nowadays consists of discarding those not considered ideal and implanting those which are, by the means mentioned earlier. But it is foreseeable that the time will come when one can not only choose from the embryos available but also that embryos can be created with those genetic characteristics which we consider desirable.

Leon Kass (see Kass 1985, 2002, p. 81) shrewdly points out that the laboratory is the door to Huxley's "brave New World" in which new human lives are manufactured. It is unanimously acknowledged that Huxley's world is a dystopia: the way society is represented is a counter-example of what human society should be. In this world the State takes exclusive charge of reproduction and does so by creating five classes of people with specific abilities to perform different tasks in society. It is a perfected form of the totalitarian eugenics that some wanted to impose in many advanced societies in the first third of the twentieth century. Nowadays there is still interest in eugenics but, according to its proponents, without the nuances which made it abhorrent in the past. It is no longer the State which controls the production of human beings, determines the most desirable genetic characteristics, or sterilises those considered unsuitable for reproduction. The aim of eugenics now is to enhance the reproductive freedom of individuals, which includes being able to choose the genetic characteristics of their progeny. If parents are looking for the best for their children throughout their life, then why not begin by choosing the best genetic characteristics? The question is whether Huxley's "brave new world" is a dystopia only because of its totalitarian context or also because of its eugenic nature. If the latter is the case, the fact that present day eugenics may be liberal and not totalitarian would not prevent the advent of a "brave new world" equally as undesirable as Huxley's, even if not totalitarian (see Agar 2004).

Since the early stages of research into DNA, the scientists most committed to the social dimension of their work discerned a hitherto unheard-of horizon in the history of humankind: for the first time human beings would be able to be the master not only the natural setting in which they live, but also their own biological makeup: "We were at an epochal moment, not only for our society or for *Homo sapiens* but for all of life on earth. For the first time in the long course of evolution, for the first time in all time, a species was coming to understand its origins and its inheritance, and with that knowledge would come the ability to alter its inheritance, to determine its own genetic destiny, as well as that of other living species. Through DNA, biology was moving beyond analysis to synthesis" (Sinsheimer 1994a, p. 135).

But many of them, including Robert L. Sinsheimer–one of the "founding fathers" of recombinant DNA and synthetic biology–,did not limit themselves to stating a new challenge for human kind. In 1966, on the occasion of the 75th anniversary of the California Institute of Technology (CalTech), Sinsheimer stated: "Ours is an age of transition. Alter two billion years, this is the end of the beginning. It would seem clear, to some achingly clear, that the world, the society, and the man of the future will be far different from that we know. Man is becoming free, not only from the external tyrannies and the caprice of toil and famine and disease, but from the very internal constraints of our animal inheritance, our physical frailties, our emotional anachronisms, our intellectual limits. We must hope for the responsibility and the wisdom and the nobility of spirit to match this ultimate freedom".[23] At this conference he mentioned some of the more attractive possibilities for human beings offered by science in general and genetics in particular: choosing sex, prolonging life, enhancing intelligence, controlling emotions, altering genetic makeup and, in short, applying intelligence to evolution.[24] But Sinsheimer does not suffice himself with simply stating these possibilities and also adopts a position. In doing so he makes clear his support for three postulates, shared by scientists and philosophers alike, from which they conceive what a human being is and should do. First, science shows that all in human beings is ultimately caused by matter.[25] Second, biotechnologies can help us to achieve the objective of our freedom more efficiently than through education.[26] Third, the State will have to extend its control over new and more intimate spheres of human life.[27]

---

[23] Sinsheimer (1966). Years later, at the time when the Asilomar moratorium was adopted in 1975, he expressed himself in the same terms: "As individuals men will have always accept their genetic constraints, but as a species we can transcend our inheritance and mould it to our purpose –if we can trust ourselves with such power. As geneticists we can continue to evolve possibilities and take the long view"; Sinsheimer (1975), p. 151. Although Jeremy Rifkin holds that Sinsheimer evolved towards more critical postures regarding the power of biotechnology, in my understanding he still retained his unfailing faith in the capacity of human beings to guide their own evolution through biotechnology; see Sinsheimer (1994b), pp. 145–146.

[24] See Sinsheimer (1966), p. 10. Although he does not cite this in his book, John Harris follows in the footsteps of Sinsheimer when giving the title to his book on human enhancement *Enhancing Evolution. The Ethical Case for Making Better People*, Princeton University Press, Princeton, 2010.

[25] "Indeed, it may be supposed that even the deepest mystery, the nature of mind and sensation and consciousness, will be understood in the end as a natural consequence of matter in a certain state of organization." Sinsheimer (1966), p. 9.

[26] "Perhaps we would like to alter the uneasy balance of our emotions. Could we be less warlike, more self-confident, more serene? Perhaps. Perhaps we shall finally achieve these long-sought goals with techniques far superior to those with which we have had to make do for many centuries;" Sinsheimer (1966), p. 10. Although not stated explicitly, he allows for thinking that education and social control might be methods that can be replaced by the superiority of biotechnology. This proposal has been taken up again recently with renewed vigour: see Douglas (2008). Opposed to the possibility of improving the moral behaviour of people through biotechnology, but from a more libertarian view in favour of any kind of enhancement of the human race, see Harris (2011).

[27] On the topic of increasing control over choosing the sex of children, Sinsheimer says: "When this prospect is combined with the already pressing problem of the expanding world population, it

Almost half a century later, all of the possibilities discerned by Sinsheimer are now the object of debate and many are interpreted as opportunities to achieve the complete liberation and enhancement of human beings. In the first decade of the twenty-first century, there have been major developments in neurosciences which have revived the conviction that we can radically enhance our intelligence, our emotions and even our moral behaviour through brain interventions. In the same way that in the 1960s, recombinant DNA was seen as the key to the coming of a new era in the evolution of humans, now it is the neurosciences that appear to have usurped this power. And today, as then, there are two opposing paradigms to interpret this new knowledge and resulting power. On the one hand, the messianic-materialist paradigm, for which the beginning and end of all human beings is material and so through manipulating matter human beings can reach their complete liberation. On the other hand, there is the pluralist paradigm, which acknowledges the material bases on which human existence is sustained –whether genetic or neurological– but rejects the postulate that a human being can be reduced to matter alone and, as a result, that the liberation of humankind lies in biotechnological manipulation of whatever kind: genetic, neurological, etc.[28]

But let us return to the possibility of altering the genetic characteristics of human beings. I shall not dwell on all the ethical reasons that have been summoned up in order to oppose them,[29] and that have been unanimously rejected almost unanimously by legal systems all over the world (see Bellver Capella 2004). I shall begin by surmising that there is a perfectly safe technology to configure the genetic makeup of each new human being, and most citizens have expressed their wish to have access to them. Two questions need to be addressed here: why is it licit to resort to this technology?; and, under what conditions should it be performed?

Without attempting to analyse all the answers that have been provided to these questions, I shall focus on the proposals of two of the leading figures who believe that germline interventions are licit (or even our duty to use), namely, the noted bioethicists John Harris and Julian Savulescu. The following analysis which focuses on their proposals attempts to illustrate that, behind apparently plausible and consistent arguments, lie hidden unresolved deficiencies and problems.

---

seems ever more clear that in the future world the right to give birth, as is today the right to take life, will have to be controlled to preserve some semblance of balance;" Sinsheimer (1966), p. 10.

[28] Adela Cortina has pointed out the radical difference between recognising the brain science bases of moral conduct, for which the neurosciences are offering priceless information, and claiming that these provide a basis to extract moral obligations, yet another attempt to reduce human beings to their material condition; see Cortina (2011).

[29] Hans Jonas and Leon Kass were the first to raise the alert concerning the risks of this possibility; see Jonas (1974). Kass (1985), pp. 43–80.

7.2.1 Why Germline Intervention Should Not Pose Moral Problems? John Harris's Answer

One of the leading present-day proponents of the right and obligation to use germline intervention to produce children with enhanced genetic qualities is John Harris.[30] His position is based on what was already defended by techno-enthusiasts in the 1970s: from the moment science offers human beings the chance to guide their own biological evolution it is their obligation to do so. Leaving aside the problems of safety and abuses, John Harris holds that germline intervention does not pose any special ethical problems because, like education, it is nothing more than an instrument to enhance our children:

"Now suppose, as is much more likely, we could use genetic engineering, regenerative medicine or drugs, or reproductive technology or nanotechnology to produce healthier, fitter and more intelligent individuals. What should reaction be? Would it be unethical to do so? Would it be ethical not to do so?

Our question is this: if the goal of enhanced intelligence, increased powers and capacities, and better health is something that we may strive to produce through education, including of course the more general health education of the community, why should we not produce these goals, if we can do so safely, through enhancement technologies or procedures?" (Harris 2010, p. 2).

The rhetorical questions posed by Harris prove extremely persuasive. But closer analysis of the reasoning behind these proposals reveals inconsistencies. The answer to Harris's question is: enhancement procedures, and in particular germline intervention, cannot be used to achieve enhanced intelligence, health or one or other capacity for the simple reason that they cannot be achieved by these means.

If they were safe, these techniques would make it possible to either combat certain exclusively genetic defects or make humans people who are different from what they have been until now. In case of the former, germline intervention does not bring happiness, guarantee health nor endow us with enhanced capacities; it simply fulfils a medical end. In the latter case, germline intervention does not bring about happiness, health and enhanced capacities either, and simply creates a human being who is different from those that have existed until now. Who can know what happiness or health is for a human who is different from us? But even if we maintain that this difference is in reality of little consequence, who can guarantee happiness, health or enhanced intelligence or other capabilities by means of manipulating genes?

When germline intervention is taken beyond a strictly therapeutic function, it engenders people who are the product of a designer. In this case human beings neither procreate nor reproduce; they are developing a product. Even if done with the best of intentions, they inevitably break with the essential symmetry between the generations as regards how everyone has been conceived until now (see Habermas 2003). The progenitor's role is replaced by the genetic "designer," who must assume responsibility for the designed product. The characteristics of the product will be

---

[30] Harris has written two books on these issues. See Harris (1992), and (2010).

unalterable. In the event that the end subject is not satisfied with the design imposed on them, can they take legal action? Under what circumstances? And who can they claim against? The designer, the person carried out the work or the person who took on the responsibility for raising him?

The aim of education is to make people better, in other words, free and happy. This aim can only be pursued by involving the freedom of the person from the very same moment it appears in his development. Without the collaboration of the liberty of the subject there can be no education. There can be training or enhancing, but not education. Therefore, it is a fallacy to invoke education to justify human enhancement.

When germline intervention is used to correct serious genetic defects we are talking about genetic medicine. When germline intervention is used to "enhance" the intelligence, health or aptitudes of future individuals, we are not doing the same we claim with education, but through other means. We are probably doing the opposite of what we do through education. This attempts to put the individual in the best of conditions to exercise her/his freedom. Germline intervention, in turn, consists of imposing on another person those characteristics the designer deems would enhance him/her, over which the subject has had no input nor can ever change. What, if anything, does this have to do with education?

It is true that education includes the development of instrumental capacities (memory, calculation, logical reasoning, etc.), some of which perhaps could end up being enhanced more effectively by germline intervention. But the aim of education is not to develop superlative instrumental capabilities in the individual for their own sake. Quite the contrary, it is only concerned with these to the extent that they are necessary for a flourishing life. While germline intervention imposes enhancements that the subject has not decided for her/himself and cannot be modified, education provides these enhancements as and when the individual wants them. And what is most important, the individual is continually redefining these enhancements.

7.2.2 How Should Germline Intervention Be Regulated. Peter Singer's Answer

The fear of Nazism is a continual reminder of attempts by the State to employ the practice of eugenics, not only in Germany during the time of the nazi movement but also in some other developed countries around the world during the first half of the twentieth century, such as the United States, Canada, Norway and Finland. In all these countries forced sterilisation programmes were approved for those people deemed unsuitable to have children (see Ridley 2006, pp. 286–300). But DNA recombinant techniques opened up an attractive eugenic panorama: select the characteristics we wanted for our children and ruled out forced sterilisations, gave us greater freedom to reproduce and ruled out State intervention, looked for ways to enhance performance of new humans and rather than sanctioning one particular race or nation (see Rifkin 1998, pp. 116 and ff). This new eugenics, labelled liberal eugenics or laissez fair eugenics (see Fitzpatrick 2001), has been defended by many scientists and philosophers, among them, Peter Singer, who backs his defence with a specific proposal to ensure that it is not counterproductive.

Singer is convinced that "citizens should choose the constitution of their government; government should not choose the constitutions of their citizens" (Singer and Wells 1984, p. 186). Consequently, he flatly rejects the idea that the State should determine the best genetic characteristics and impose germline intervention on its citizens to endow their progeny accordingly. Neither does he consider it correct to simply leave the genetic selection of future generations in the hands of the market because, "it puts too much power in the hands of individuals who might use it irresponsible or even pathologically" (Singer and Wells 1984, p. 186). Singer holds that "the genetic endowment of children should be in the same hands it always have been – the hands of parents." But if the parents wished to, for the first time, to incorporate a genetic characteristic in their children by means of germline interventions, they should require the pertinent authorisation from a public body created for this purpose. "A broadly based government body could be set up to approve or reject particular parents' proposals for genetic engineering. It would consider whether the proposed piece of engineering would, if its practice became widespread, have harmful effects on individuals or society. If no harmful effects could be foreseen, the committee would license the procedure. This would mean that parents who wish to use it were free to do so" (Singer and Wells 1984, p. 188).

Singer's proposal has many valuable points in its favour. For extremely complex and constantly changing areas such as biotechnology, it is much more operative to entrust a professional multidisciplinary and plural body with the power to propose regulations or authorise, or not, certain practices after having studied each case. This has been adopted in certain areas of bioethics; countries such as the United Kingdom or Spain have State bodies authorised to govern practices or experiments in certain areas of biomedicine. The best known example is the Human Fertilization and Embryo Authority (HFEA).

However, if there were a body responsible for authorising germline intervention for future parents, which according to their criteria would not represent a danger either for individuals or society, then would it work? To my way of thinking such a proposal is unviable for several reasons:

– It is impossible for a commission of this nature to come to a broad agreement if it attempts to reflect the various postures adopted on germline intervention in society. To be truly operative its members must share the idea that germline intervention is positive providing it does not lead to abuse. Then it would be able to discuss whether a certain genetic "enhancement" involves a risk or not for the individual or society. But then these commissions would no longer reflect the plurality of views in society: one group would be imposing its particular view of what is good over the others.

– In a globalised world it is impossible for a commission with these characteristics to be effective in real terms if the area in which it operates is not universal. It would be able to prevent the practice of certain types of germline intervention, but it could not prevent the citizens from going to another country where they are authorised, should they so wish. Reproduction tourism, which at present is limited to the search for certain assisted reproduction services when they are prohibited in their own country or prohibitively costly, will extend to germline intervention by necessity.

When the market is global, national rulings on what should be the subject of strict regulation are not enough to protect against the law of supply and demand. Any restriction on access to germline intervention established by a leading country in biotechnology will be taken advantage of by others to attract investments, researchers, and customers eager to gain access to what is banned in their own countries.

7.3 Conclusions

Three main conclusions can be drawn in the light of what has been discussed in this chapter:

– Biotechnologies are developing at present in a scenario characterised by three profoundly new elements with respect to what has been scientific and technological development until the present day. Firstly, biotechnologies form part of a conglomerate of relations made up of public powers, private enterprise and citizens, which determine their development entirely. Secondly, biotechnologies depend on increasingly greater financial resources, and obtaining these determines and orients them. Thirdly, alongside increasing financial dependence, biotechnologies are also experiencing decreasing dependence on laws and regulations bringing undesired effects in its wake.

– Biotechnologies are moving towards total intervention in human biology, as has already happened in non-human biology. This route began with the use of assisted reproduction techniques on humans and is frequently legitimised arguing that these uses of biotechnology can lead to creating better humans (human enhancement).

– Those authors who defend the right, or even the obligation of human enhancement by means of biotechnology use arguments they themselves would not accept if they were to subject them to rigorous analysis with which they judge opposing arguments to human enhancement, which they often qualify contemptuously as intuitive.

References

Agar, N. 2004. *Liberal eugenics. In defence of human enhancement.* Oxford: Blackwell.
Annas, G. 2011. Assisted reproduction — Canada's supreme court and the 'global baby'. *The New England Journal of Medicine* 365: 459–463.
Ballesteros, J. 2007. Biotecnología, biopolítica y posthumanismo. In *Biotecnología y posthumanismo*, ed. J. Ballesteros and E. Fernández. Cizur Menor: Thomson-Aranzadi.

Bellver Capella, V. 2004. Las intervenciones genéticas en la línea germinal humana y el horizonte de un futuro posthumano. In *Biotecnología, dignidad humana y derecho: Bases para un diálogo*, ed. J. Ballesteros and A. Aparisi, 115–148. Pamplona: EUNSA.

Bellver Capella, V. 2012. Embriones humanos clónicos triploides: Aspectos éticos, sociales y jurídicos. *Revista de Derecho y Genoma Humano* 36: 25–64.

Bostrom, N., and J. Savulescu (eds.). 2010. *Human enhancement*. Oxford: Oxford University Press.

Buyx, A.M. 2008. Be careful what you wish for? Theoretical and ethical aspects of wish-fulfilling medicine. *Medicine, Health Care and Philosophy* 11(2): 133–143.

Cortina, A. 2011. *Neuroética y Neuropolítica. Sugerencias para la educación moral*. Madrid: Tecnos.

Davo, M.C., and C. Alvarez Dardet. 2003. El Genoma y sus metáforas: ¿Detectives, héroes o profetas? *Gaceta Sanitaria* 17(1): 59–65.

Douglas, T. 2008. Moral enhancement. *Journal of Applied Philosophy* 25(3): 228–245.

Editorial. 2007. Behind the stem cell breakthrough. *The New York Times*, December 1. http://www.nytimes.com/2007/12/01/opinion/01sat1.html?_r=0. Accessed on 23 Jan 2012.

Editorial. 2010. The human genome at ten. *Nature* 464: 649–650.

Fitzpatrick, T. 2001. Before the cradle: New genetics, biopolicy and regulated eugenics. *Journal of Social Policy* 30(4): 589–612.

González Quirós, J.L., and J.L. Puerta. 2009. Tecnología, demanda social y medicina del deseo. *Medicina Clínica* 133(17): 671–676.

Habermas, J. 2003. *The future of human nature*. Oxford: Blackwell/Oxford.

Harris, J. 1992. *Wonderwoman and superman. The ethics of human biotechnology*. Oxford: Oxford University Press.

Harris, J. 2010. *Enhancing evolution. The ethical case for making better people*. Princeton: Princeton University Press.

Harris, J. 2011. Moral enhancement and freedom. *Bioethics* 25(2): 102–111.

Human Fertility and Embryology Authority (press release), HFEA agrees new policies to improve sperm and egg donation services. http://www.hfea.gov.uk/6700.html. Accessed on 1 Feb 2012.

Hwang, W.S., et al. 2004. Evidence of a pluripotent human embryonic stem cell line derived from a cloned blastocyst. *Science* 303: 1669–1674.

Hwang, W.S., et al. 2005. Patient-specific embryonic stem cells derived from human SCNT blastocysts. *Science* 308: 1777–1783.

Inhorn, M., and P. Patrizio. 2009. Rethinking reproductive 'tourism' as reproductive 'exile'. *Fertility and Sterility* 92(3): 904–906.

Jonas, H. 1974. Biological engineering: A preview. In *Philosophical essays: From ancient creed to technological man*, ed. H. Jonas, 141–167. Englewood Cliffs: Prentice Hall.

Kass, L.R. 1985. *Toward a more natural science. Biology and public affairs*. New York: The Free Press.

Kass, L.R. 2002. *Life, liberty and the defense of dignity*. San Francisco: Encounter.

Kennedy, D. 2001. Legislate in haste, repent at leisure. *Science* 294: 745.

Levitt, M., and H. Zwart. 2009. Bioethics: An export product? Reflections on hands-on involvement in exploring the 'external' validity of international bioethical declarations. *Bioethical Inquiry* 6: 367–377.

Marshall, E. 2011. Waiting for the revolution. *Science* 331: 525–528.

Mori, M., and D. Neri. 2001. Perils and deficiencies of the European convention on human rights and biomedicine. *Journal of Medicine and Philosophy* 26(3): 323–333.

Mouzon, J., et al. 2010. Assisted reproductive technology in Europe, 2006: Results generated from European registers by ESHRE. *Human Reproduction* 25: 851–1862.

Nielsen, T.H. 2008. What happened to the stem cells. *Journal of Medical Ethics* 34: 852–857.

Nisbet, M., D. Brossard, and A. Kroepsch. 2003. The stem cells controversy in an age of press/politics. *Press/Politics* 2(8): 36–70.

Noggle, S., et al. 2011. Human oocytes reprogram somatic cells to a pluripotent state. *Nature* 478: 70–75.

Pennings, G. 2004. Legal harmonization and reproductive tourism in Europe. *Human Reproduction* 19(12): 2688–2692.

Pollack, A. 2012. Stem cell treatment for eye diseases shows promise. *The New York Times* January 23. Available in: http://mobile.nytimes.com/2012/01/24/business/stem-cell-study-may-show-advance.html. Accessed on 23 Jan 2013.

Ridley, M. 2006. *Genome. The autobiography of a species in 23 chapters*. New York: Harper-Perennial. Genome.

Rifkin, J. 1998. *The biotech century. Harnessing the gene and remaking the world*. New York: Penguin.

Sánchez Ron, M. 2000. *El siglo de la ciencia*. Madrid: Taurus.

Schuklenk, U. 2010. Defending the indefensible. The UNESCO declaration on bioethics and human rights: A reply to Levitt and Zwart. *Bioethical Inquiry* 7: 83–88.

Schwartz, S. et al. 2012. Embryonic stem cell trials for macular degeneration: A preliminary report. *The Lancet* 379(9817):713–720. doi:10.1016/S0140-6736(12)60028-2. http://download.thelancet.com/flatcontentassets/pdfs/S0140673612600282.pdf. Consulted on 28 Jan 2012.

Singer, P., and D. Wells. 1984. *The reproduction revolution. New ways of making babies*. Oxford: Oxford University Press.

Sinsheimer, R. 1966. The end of the beginning. *Engineering and Science* 30(3): 7–10.

Sinsheimer, R. 1975. Troubled dawn for genetic engineering. *New Scienstist* 68(971): 148–151.

Sinsheimer, R. 1994a. *The strands of a life: The science of DNA and the art of education*. Berkeley: University of California Press.

Sinsheimer, R. 1994b. The prospect of designed genetic change. In *Ethics, reproduction and genetic control*, ed. R. Chadwick, 136–146. New York: Routledge.

Spar, D. 2006. *The baby business how money, science, and politics drive the commerce of conception*. Boston: Harvard Business School Press.

Stolberg, S.G. 2001. Despite warnings, 3 vow to go ahead on human cloning. *The New York Times*, August 9. Available in: http://www.nytimes.com/learning/teachers/featured_articles/20010809thursday.html. Accessed on 23 Jan 2013.

Varmus, H. 2010. Ten years on – The human genome and medicine. *The New England Journal of Medicine* 362: 2028–2029.

Vogel, G. 1999. Capturing the promise of youth. *Science* 286: 2238–2239.

Vogelstein, B., B. Albert, and K. Shine. 2002. Please, Don't call it cloning! *Science* 295: 1237.

Part III
Technology and Risks

Chapter 8
Risk and Trust in Institutions That Regulate Strategic Technological Innovations: Challenges for a Socially Legitimate Risk Analysis

Hannot Rodríguez

8.1 Theoretical Framework

Technological innovation has become the cornerstone of economic growth and competitiveness. However, as such innovation sometimes generates new risks to health and the environment, it needs to be considered not only as a source of wealth and progress, but also as a potential source of environmental and health hazards.

Regulatory institutions took risk analysis on board in a bid to make science and technology development safer; it was also seen as a means of legitimizing such development. But techno-industrial progress frequently provokes resistance in society, which sees such progress as dangerous, despite the fact that institutional risk analysis concludes risk levels are acceptable. Institutions controlling progress have often played down such reactions, describing them as irrational or prejudiced. Nevertheless, over the last couple of decades, institutions have gradually come to admit the relevance and legitimacy of societal concerns and the distrust of risk analysis and technological progress. Similarly, social studies of risk have dignified social resistance to technological progress through specific research into the relations of trust between publics and the institutions responsible for safety.

Increasing societal resistance to technological innovations, such as the public backlash in Europe against agri-food biotechnology, has prompted regulatory institutions to introduce major modifications in risk analysis which, as "risk governance," has become more participative, more precautionary, and more sensitive to the ecological and socio-ethical dimensions of decisions, partly in an attempt to re-legitimize their activity and, consequently, guarantee the feasibility of a society

H. Rodríguez (✉)
Department of Philosophy, Faculty of Arts, University of the Basque Country,
Paseo de la Universidad, n. 5, Vitoria 01.006, Spain
e-mail: hannot.rodriguez@ehu.es

© Springer International Publishing Switzerland 2015   147
W.J. Gonzalez (ed.), *New Perspectives on Technology, Values,*
and Ethics, Boston Studies in the Philosophy and History of Science 315,
DOI 10.1007/978-3-319-21870-0_8

based to a great extent on progress in science and technology. However, significant socio-economic constraints on risk governance limit its ability to conceive and promote alternative safety scenarios and more socially robust, or legitimate, decisions on risks.

In this paper I examine the challenges facing regulatory institutions in Europe concerning risk analysis of strategic technological innovations such as agri-food biotechnology and nanotechnology, and the legitimacy issues these challenges raise. To this end, I argue first that dynamics of trust and distrust between publics and regulatory institutions respond to legitimate concerns, or challenges, that affect the principles and capabilities of institutional risk analysis. Next, based on a previous study (Rodríguez 2009), I analyze these challenges, presenting three models for the understanding of trust: the competence model, the cultural model and the relational model. I argue that each of these models refers to a specific challenge that institutional risk analysis in our societies has to overcome if it is to be accepted as legitimate: an epistemological challenge, an axiological challenge and a reflexive challenge. I illustrate these challenges mostly in the light of the controversial regulation of transgenic plants in Europe. Then I argue that risk governance measures such as the adoption of precautionary policies and the promotion of public participation exercises, designed to resolve and avoid both the challenges and the partial deficit of legitimacy affecting regulatory institutions and technological developments, are severely conditioned by the strategic socio-economic imperatives that guide scientific-technological innovations and regulatory policies. I use the European pro- motion and regulation of nanotechnology to illustrate the constraints that severely limit institutional capacity and willingness to develop alternative and potentially more socially legitimate techno-industrial safety scenarios. Finally, I present some general conclusions.

8.2   The Value of Social Trust in Risk Analysis

Technological innovation has become the cornerstone of economic growth and competitiveness. The advanced economies of the world (e.g. Europe and the United States) cope with economic difficulties in both domestic and international arenas, such as high unemployment, outsourcing of production, and emergent economies. They have dramatically increased funding and support for science and technology research and development programs as a way to gain an innovative edge in a highly competitive international market (Biegelbauer and Borrás 2003; Marklund et al. 2009).

However, technological innovation can bring new risks to health and the environment, and needs to be considered not only as a source of wealth and progress, but also as a potential source of environmental and health hazards (Cranor 2011; Shrader- Frechette 2011). Industrialized countries use risk analysis to make science and technology development safer. Risk analysis involves the scientific assessment and political management of risks: scientific assessment identifies and quantifies risks, while risk management develops and implements the necessary political-legal measures to control them, based on the knowledge provided by risk assessment (Shrader-Frechette 1991).

But the goal of risk analysis is not solely to offer safer technological developments. It also aims to legitimize techno-industrial progress. Risk analysis was established in the late 1960s, in the context of radical social criticism of industrial society (Dickson 1984, pp. 261–306). Instead of promoting a profound revision of the basic political, scientific-technological and economic dynamics of modern societies, institutions established risk analysis, on the assumption that the industrial development of science and technology could be controlled without renouncing the basic premises of social progress (productivity, economic growth, employment, competitiveness) (Luján and López Cerezo 2004).

Nevertheless, the legitimizing capacity of risk analysis in our societies is not unlimited. Certain social groups frequently criticize and resist technological innovations on the grounds that they are dangerous, despite institutional risk analysis concluding that the risk levels involved are acceptable. A good illustration of this is the backlash in Europe against agri-food biotechnology (Bauer and Gaskell 2002; Gaskell 2008).

Initially, institutions responsible for safety interpreted social reactions against scientific and technological developments as irrational responses based on a lack of understanding of real risks. In the 1980s, in a context of strong social resistance to nuclear power, institutions decided to add a third function to risk analysis: risk communication, the assumption being that the diffusion of "objective" information about risks to society would facilitate societal acceptance of technologies (Charnley et al. 2000, pp. 304–305).[1]

The effectiveness of this strategy has, however, been limited. The European Commission acknowledged that risk analysis "has become a crucial but often highly controversial component of public policy" (European Commission 2002, p. 23), and this should come as no surprise: in a knowledge-based society that relies on fast-growing technological innovations, where possible negative impacts are often uncertain, the economic stakes are high and social sensitivity to environmental and health protection is constantly increasing, controversy over safety measures are arguably normal (Ravetz 2003).

Different publics may perceive decisions as impositions, designed, among other things, to favor industrial interests; they may also feel risk is not fairly distributed, or that risk assessments are biased and uncertainties underestimated (Pellizzoni 2001). In other words, public attitudes toward risk cannot be dismissed as a mere result of risk misrepresentation; they are based on judgments about institutional behavior and commitments, in a context where science- and technology-related

---

[1] The assumption that societal reactions against scientific-technological progress are the result of irrational appraisals of risk continues to pervade more contemporary institutional responses to these attitudes. For instance, public reticence in Europe about accepting transgenic food was considered by the then European Commissioner for Health and Consumer Protection David Byrne as something "inconsistent if not completely irrational," Byrne (2003), 2.

decisions imply significant socio-economic, political and ecological consequences (Cobb and Macoubrie 2004, p. 395; Hansson 2005, p. 79). The publics therefore may simply not trust institutional ability to govern techno-industrial risks in a satisfactory way.

Dynamics of societal trust and mistrust toward science, technology and expert regulatory institutions thus respond to legitimate problems and concerns about the development, implementation and regulation of science- and technology-based industrial activities, and not simply to groundless irrational preoccupations (e.g. Slovic 1999). In modern industrial "risk societies," the major institutions (science, economy, politics, law) arguably have limited capacity to deal with techno-industrial, or "manufactured," risks. Here, irresponsibility is not related to irrational publics, but to the very structure and normal dynamics of industrial societies, which are relatively unable to keep techno-industrial risk under control (Beck 1986 [1992], 1988 [1995]). Recent accidents such as the March 2011 Fukushima Daiichi nuclear disaster in Japan or the BP oil spill in the Gulf of Mexico in April 2010 highlight the limitations of human control capabilities in complex socio-technical systems (Ebert 2012). Together with accumulative, or "chronic" dangers such as climate change, or industrially induced global warming (Giddens 2009), the emergence of these risks may be interpreted as a fundamental and characteristic anomaly of our modern knowledge-based industrial societies.

The authority of modern institutions has been curtailed by challenges posed by the risks and uncertainties of progress. The food crises suffered by Europe in the 1990s (i.e., "mad cow" disease, foot and mouth disease, and dioxin contamination in chickens) underscored the limitations of risk analysis and created a feeling in society that policymakers were doing more to favor industry than the public interest, which, as the European Commission admitted, "undermined public confidence in expert-based policy-making" (Commission of the European Communities 2001, p. 19). Similarly, social studies of risk—i.e., investigations that analyze the epistemological, political and cultural principles and dynamics underlying risk analysis and risk perception—have examined public trust and distrust of regulatory institutions based on significant and legitimate challenges, and have accordingly dignified the status of skeptical and critical public reactions in the sense that they are claimed to be related to a series of substantive challenges to risk analysis and techno-economic progress, and not just to failed and/or "symbolic"[2] representations of risks (e.g. Jacob and Hellström 2000; Wynne 2006).

---

[2] Critical public perceptions and attitudes to technological developments have been traditionally considered cognitively wrong, or failed, representations, and also as intentionally impartial, or "symbolic," risk appraisals. For instance, antinuclear groups have often been said to use the debate on the risks of nuclear power "as a surrogate for larger policy questions about desired life-styles, political structure (…), and institutional power," where "evidence about actual impacts is almost meaningless for the actors, but is still a desired resource to mobilize support" (Renn 1992, 191). But the political significance and meaning of the risks of progress is not merely symbolic-conventional; as I have already said, risk is a constitutive characteristic of modern techno-industrial societies and, therefore, its relation with the socio-political dimension is not conventional, but *meaningful* (e.g. Beck 1986 [1992]).

Dynamics of public trust and distrust between publics and expert institutions need, therefore, to be understood in the context of substantive and legitimate challenges. In the following section I analyze these challenges, presenting three models for the understanding of trust: the competence model, the cultural model and the relational model. I argue that each of these models refers to an essential challenge that institutional risk analysis in our societies has to deal with in order to be seen as legitimate: an epistemological challenge, an axiological challenge and a reflexive challenge. I illustrate these challenges mostly in the light of the controversial regulation of transgenic plants in Europe.

8.3 Three Models of Trust: Challenges for a Socially Legitimate Risk Analysis

Trust can be defined as a positive expectation regarding the behavior of the other in a risky situation (Das and Teng 2001, p. 255). However, this general characterization of trust does not specify how optimistic expectation (as well as its absence) is constituted. The three models of trust presented here (the competence model, the cultural model and the relational model) offer alternative interpretations of how trust is placed in institutions responsible for safety, in the sense that these models point to, respectively, an epistemological, an axiological and a reflexive challenge involved in trust and distrust dynamics between publics and expert institutions.

8.3.1 The Competence Model of Trust and the Epistemological Challenge

According to the competence model, public trust in regulatory institutions depends on the level of competence these institutions demonstrate in risk control.

The main champion of this model, Anthony Giddens, claims that the functioning of modern institutions depends to a great extent on *active* trust (Giddens 1990, p. 26). What this means is that, unlike traditional societies, modern institutions are not legitimized in absolute terms: the validity and legitimacy of modern institutions are under revision (Giddens 1994a, p. 89).

Giddens differentiates between "confidence" and "trust." The first corresponds to pre-modern societies; the second is a feature of modern societies. Confidence relates to an unquestionable normative framework, while trust is active confidence (Giddens 1990, 1994a, pp. 85–91). He makes an additional distinction between simple modernity and reflexive modernity, a distinction that also marks the transition between the two types of confidence (Giddens 1990, 1994a). In simple modern times (until approximately four decades ago), the authority of science in terms of competence was unchallenged, and science worked as a tradition in modern societies, even though science is a propositional system of knowledge based on revisability (Giddens 2003, pp. 31–35).

In Giddens' view, science loses its absolute legitimacy status in the eyes of society once the publics become aware of the limitations of expert knowledge in the determination of scientific-technological risks (Giddens 1990, pp. 124–131, 2003, pp. 25–26). This awareness is based on the social perception of expert disagreement on risk (Giddens 1994b, pp. 185–186, 2003, pp. 31–32). Public skepticism toward science is thus the consequence of scientific skepticism. Public awareness of uncertainty causes the transition from trust as confidence to trust as active confidence:

"The faith that supports trust in expert systems involves a blocking off of the ignorance of the lay person when faced with the claims of expertise; but realisation of the areas of ignorance which confront the experts themselves, as individual practitioners and in terms of overall fields of knowledge, may weaken or undermine that faith on the part of lay individuals. Experts often take risks "on behalf" of lay clients while concealing, or fudging over, the true nature of those risks or even the fact that there are risks at all. More damaging than the lay discovery of this kind of concealment is the circumstance where the full extent of a particular set of dangers and the risks associated with them is not realised by the experts. For in this case what is in question is not only the limits of, or the gaps in, expert knowledge, but an inadequacy which compromises the very idea of expertise" (Giddens 1990, pp. 130–131).

Therefore: (i) skeptical oscillation of public trust toward science is triggered by the social awareness of expert uncertainty; (ii) ignorance is a necessary condition for the formation of trust judgments (Giddens 1990, p. 89); (iii) the public makes up for its lack of knowledge or competence to judge expert systems with an act of *faith* (Giddens 1990, p. 34). Trust, then, is a confidence that expresses "a faith in the probity or love of another, or in the correctness of abstract principles (technical knowledge)" (Giddens 1990, p. 34).[3]

As we have already seen, expert disagreement about risks would weaken that faith and, in consequence, provoke the oscillation of public trust in expert systems. For instance, despite the fact that transgenic Bt[4] maize MON 810 from Monsanto has authorization to be grown for commercial purposes in the European Union,[5] nine EU countries ban the cultivation of MON 810 in their national territories,[6]

---

[3] This does not mean that trust in expert institutions can be reduced completely to a state of faith. Trust must also be based on the appropriation of the processes through which risk is socialized and becomes acceptable: regulatory measures, experiences of previous adequate operation, educational processes that socialize the respect toward science and technology, etc. (Giddens 1990, 35, 88–92).

[4] A Bt crop is one that expresses proteins with insecticide properties produced by the bacteria *Bacillus thuringiensis* (Bt). See, for instance: U.S. Environmental Protection Agency, "EPA's Regulation of Bacillus thuringiensis (Bt) Crops," 735-F-02-013 (May 2002), www.epa.gov/oppbppd1/biopesticides/pips/regofbtcrops.htm (Accessed on February 27, 2012).

[5] D. Butler 2010, "A new dawn for transgenic crops in Europe?", *Nature News* (March 9, 2010, doi:10.1038/news.2010.112), www.nature.com/news/2010/100309/full/news.2010.112.html (Accessed on February 27, 2012).

[6] The countries are Austria, Hungary, Greece, France, Luxembourg, Italy, Germany, Bulgaria and Poland. See: Greenpeace Poland, "Genetically modified crops illegal – Government launches bans," in: F. Kreiss (Occupy Monsanto), *Poland is the Most Recent Country to Ban GMO*

based on what they argue is scientific evidence of environmental risk, evidence provided by their national scientific committees. For example, one of these countries, France, informed the other Member States about the ban as follows: "Given this new scientific evidence, the French authorities considered that the cultivation of MON 810 maize was liable to present a serious threat to the environment" (République Française 2009, p. 2).[7] By contrast, the European Food Safety Authority (EFSA), consulted by the European Commission, concluded that the French Government had shown no evidence of the environmental and health risks of this maize: "there is no specific scientific evidence, in terms of risk to human and animal health and the environment" (Panel on Genetically Modified Organisms 2008, p. 31).[8] This disagreement about the evidence of the safety of genetically modified organisms (GMOs), as argued in the competence model, seems to legitimize a partially mistrustful public attitude toward expert regulatory institutions, given that it points to a fundamental challenge for the anticipatory scientific assessment of risk analysis and its promises of knowledge and control.

8.3.2   The Cultural Model of Trust and the Axiological Challenge

Unlike the previous model, the cultural model denies that trust in regulatory institutions depends on competence criteria and objective risks. The founders of this perspective, Mary Douglas and Aaron Wildavsky, emphasize the cultural, or contextual, character of all risk experiences (Douglas and Wildavsky 1982). In their opinion, the significance of threat is not given, but socioculturally constituted.

More exactly, risk here is understood as a resource that each social group in society uses for its own benefit (Douglas and Wildavsky 1982, pp. 29–48). These authors identify three social groups: the hierarchist or bureaucratic, the market individualist, and the sectarian. The first two represent society's status quo or what they call the "center" (Douglas and Wildavsky 1982, pp. 83–101), and the third one represents the anti-system branch of society, i.e., the "border" (Douglas and Wildavsky 1982, pp. 102–151).

Cultivation (January 19, 2013), www.occupymonsanto360.org/2013/01/19/poland-is-the-most-recent-country-to-ban-gmo-cultivation (Accessed on March 29, 2013).

[7] The French experts claimed evidence of the following environmental risks of MON 810: (i) the environmental dissemination, contamination and persistence of Bt toxin, (ii) the appearance of resistance strains in target pests, and (iii) the development of toxic traces in non-target fauna (Comité de préfiguration d'une haute autorité sur les organismes génétiquement modifiés 2008, 1–2).

[8] The French ban on the cultivation of the MON 810 was overturned by the country's top court in November 2011 on the basis that it was not sufficiently justified, but the Government reinstated the ban in March 2012. See: S. de La Hamaide (Reuters), "France restores ban on GMO maize crops" (March 16, 2012), www.reuters.com/article/2012/03/16/us-france-gmo-idUSBRE82F16I20120316 (Accessed on March 19, 2013).

The different social groups are not interested in the same risks. The social border, for example, appeals to technological, or ecological, risks in order to attack what this border considers to be immoral (*polluted*) forms of economic and political power, namely the social center (Douglas and Wildavsky 1982, p. 47).

In this model, the trust and mistrust dynamics precede opinion on the risks, in contrast to what the competence model claims. In other words, the competence of regulatory institutions cannot be neutrally assessed, an argument reiterated in further developments of the cultural model (e.g. Earle and Cvetkovich 1999, pp. 20–21). The placement of trust is based here on the *similarity between cultural values*, meaning that "people tend to trust other people and institutions that 'tell stories' expressing currently salient[9] values, stories that interpret the world in the same way they do" (Earle and Cvetkovich 1999, p. 11; see also: Cvetkovich 1999; Earle and Cvetkovich 1999; Earle et al. 2007). Trust placements and the subsequent risk perception are thus conditioned by "social identity" (Cvetkovich and Winter 2007, pp. 192–193). For example, environmental organizations are highly valued in society because of their ability to show themselves as the alternative to industrial pollution and depredation (Siegrist et al. 2007).[10]

The competence and cultural models do agree on something, however. Social trust in regulatory institutions is based on the fact that the publics are not capable of "rationally" assessing how competent expert systems are in controlling risks (Cvetkovich 1999, p. 59). Nevertheless, in the cultural model this inability is not addressed through a "faith (…) in the correctness of abstract principles (technical knowledge)" (Giddens 1990, p. 34), but through an impression of shared values (Johnson 2007, p. 211 and 235). Thus, the public management of ignorance is not monitored here through faith in the efficiency of science, as the competence model argues, but through a series of perceptions focusing on the socially most valued problems.

Thus, according to this model, technological proposals are debated not so much in regard to the magnitude of their risks and uncertainties, but in regard to the manner — i.e., the principles — according to which those proposals are promoted. For instance, one of the factors underlying the early societal resistance in Europe to GMOs in the 1990s was the lack of labeling of transgenic products, since the original regulatory Directive 90/220/EEC did not demand specific labeling of genetically modified products (The Council of the European Communities 1990). Consumers were therefore denied their right to choose to consume transgenic products or not (Winickoff et al. 2005, pp. 87–88), which gave rise to several boycott campaigns (Levidow and Murphy 2003, p. 63). As a result, authorities imposed a de facto

---

[9] They are "salient" because the set of values linked to a given individual is not fixed but can vary according to the shift of meanings of the circumstances that are being valued. Douglas and Wildavsky's proposal, where individuals are subsumed under a characteristic set of values that monopolize their opinions, is thus diluted here (Siegrist et al. 2000, 355).

[10] Siegrist et al. (2007) distinguish between trust based on facts (*confidence*) and trust based on values (*trust*). In their opinion, institutional strategies that try to obtain public legitimacy only through confidence are doomed to fail.

moratorium on transgenic products between 1998 and 2004 (Winickoff et al. 2005, pp. 88–90), and in the meantime developed stricter normative frameworks that imposed labeling and traceability conditions on all foodstuffs (Todt 2004, p. 150),[11] by which the moratorium on new GMOs was lifted in Europe.[12] Transgenic products continue to be controversial in Europe, but the experience of their introduction there shows us that the legitimacy of decisions about technology safety in our societies cannot rest exclusively on expert and technical criteria; fundamental value principles of democratic societies, such as the right to information and choice, play a major role, as the cultural model propounds, in the constitution and maintenance of public trust in institutions.

8.3.3   The Relational Model of Trust and the Reflexive Challenge

Both the competence model and the cultural model consider trust expectations to be voluntary. For example, Mary Douglas, one of the precursors of the cultural model, claims that "the individual makes an initial choice for a kind of organization and this commitment itself generates the decision-making and perceptual bias" (Douglas 1985, p. 89). The relational model, instead, understands trust as a phenomenon associated with the power relationships of society. According to Brian Wynne, champion of this perspective, "trust and credibility are contingent variables which depend upon evolving relationships and identities" (Wynne 1996a, p. 20).

On the other hand, this model discards the idea of neutral science, and claims that scientific knowledge is based on tacit cultural assumptions such as predictability and controllability (Wynne 1992). The relational model, then, erodes the demarcation, reproduced by the two previous models, between facts (science) and values (culture). Science, which is no longer neutral knowledge, imposes its culture on society (Wynne 1996b, pp. 57–60 and 67–68), and this circumstance conditions the dynamics of social trust in expert institutions.

Wynne illustrates this idea with the example of the expert management of Chernobyl post-accident contamination levels in the region of Cumbria (north of England). This management was characterized by strong opposition from farmers to the experts' conclusions and measures, which ended up being ineffective and unrealistic (Wynne 1996a, b, pp. 62–68). Wynne argues that this happened because

---

[11] The new, stricter regulations were Directive 2001/18/EC from 2001 (The European Parliament and the Council of the European Union 2001), which repealed the previous Directive 90/220/EEC, and the Regulations 1829/2003 (The European Parliament and the Council of the European Union 2003a) and 1830/2003 (The European Parliament and the Council of the European Union 2003b), both from 2003.

[12] However, many supermarkets continue to refrain from adding genetically modified ingredients to their brand products. For example, most distribution chains in Spain follow this policy. Moreover, a large number of producers have renounced the use of transgenic organisms in their products (Greenpeace 2012).

the cultural premises of controllability by the experts were unreflexively imposed on local knowledge and practices, which were more prone to accept uncertainty (Wynne 1996a, pp. 34–37, b, pp. 66–67). In his own words:

"The farmers assumed predictability to be intrinsically unreliable as an assumption, and therefore valued adaptability and flexibility, as a key part of their cultural identity and practical knowledge. (…) the two knowledge-cultures expressed different assumptions about agency and control (…)" (Wynne 1996b, p. 67).

Wynne is describing a conflict between identities, where mistrust in expert institutions is a result of the imposition of unrealistic control claims on the publics, who perceive these claims as alien cultural principles, i.e., as "identity-risks" (Wynne 1996a, pp. 55 and 59–60, b, pp. 35–36 and 39).[13] Unlike Giddens, Wynne argues that mistrust does not derive from expert disagreement (i.e., from uncertainty), but from the opposite: excessive trust and certainty in the scientific capacity to control risks (Wynne 2006, pp. 215–217).

According to this model, neither uncertainty (competence model) nor the misalignment between subjective values unrelated to science (cultural model) explain the emergence of social mistrust toward regulatory institutions. Social mistrust is related here to the unreflexive and overconfident culture guiding scientific practices on risk. For instance, the original Directive 90/220/EEC on the deliberate release into the environment of genetically modified organisms did not acknowledge the systemic uncertainties surrounding agri-food biotechnology applications and, therefore, assumed that the environmental risks associated with transgenic crops could be adequately identified and assessed (The Council of the European Communities 1990). However, science does, for instance, calculate risks of transgenic introgression[14] through indirect scientific methods: in the absence of direct scientific evidence, the probability of gene introgression is extrapolated from knowledge about sexual compatibility between crops and wild relatives, and information obtained from areas of geographic proximity between crops and wild relatives (Murphy and Krimsky 2003, pp. 135–136). Aware of this limitation, once impelled by the societal concern and controversy over the risks of GMOs, the Directive 2001/18/EC, which repealed the previous Directive 90/220/EEC, states the obligation to implement a post-commercialization monitoring plan for authorized GMOs "in

---

[13] The subordination of publics to expert institutions can conceal public unrest and mistrust toward them: due to this dependence on institutions, publics act "as if" they trust institutions, although in fact they do not. This circumstance highlights, according to this model, the socio-relational and *ambivalent* character of the relations between science and the publics (Wynne 1996a, 40–42, b, 50–52, 65, 68). In this sense, the adequacy of calling "trust" a forced expectation could be denied. It has been argued that in this case we should talk about "compliance" rather than "trust" (Möllering 2006, 119). However, we will assume, as Wynne does, that it is correct to talk of "trust" under circumstances of social coercion in regard to what the relational model of trust reveals: the latent mistrust hidden under an apparent public legitimacy.

[14] Introgression is the movement of genes from one species to another or among sub-species that have been geographically isolated. See: C. Maynard 1996, "Forest Genetics Glossary" (*State University of New York College of Environmental Science and Forestry*), www.esf.edu/for/maynard/GENE_GLOSSERY.html (Accessed on March 19, 2013).

order to trace and identify any direct or indirect, immediate, delayed or unforeseen effects on human health or the environment of GMOs as or in products" (The European Parliament and the Council of the European Union 2001, p. 3). This reflects an explicit acknowledgment of the systemic uncertainties surrounding agri-food biotechnology applications, and that the only way to assess the safety of these transgenic technologies is to implement them in the real world (Todt 2002, pp. 101–103). According to the relational model of trust, this explicit acknowledgment of the limitations of science together with the reinforcement of the control measures of GMO products should improve trust in regulatory institutions.

8.4 Going Through the Challenges? Risk Governance and Its Limits

The epistemological, axiological and reflexive challenges associated with the dynamics of trust and distrust between publics and the institutions that regulate technological risk highlight the fact that the effectiveness and social legitimacy of regulatory institutions depend on a variety of epistemic, ethical and socio-political dimensions and their interactions. Safety scenarios in our societies are decided and implemented according to available knowledge, guiding principles, socio-political priorities, goals and power dynamics, and to the ways in which all these relate. As we have seen in the regulation of transgenic organisms in Europe, where regulatory institutions were forced to promote some regulatory adjustments such as the acknowledgement of the consumer's right to choose and the imposition of stricter, more precautionary scientific monitoring measures, the societal and instrumental robustness of risk analysis depends on an integral and reflexive consideration and analysis of a variety of socio-scientific issues.

These regulatory adjustments were conceived, developed and implemented as ad hoc measures in response to public reticence about, and resistance to transgenic technologies. However, in the last two decades, the reforms have become increasingly stable in European risk analysis. In part as an attempt to re-legitimize their activity and, consequently, guarantee the feasibility of a society based to a large extent on progress in science and technology, regulatory institutions have introduced important changes in the analysis of risk which, as risk "governance," has become more participative, more prudent (by means of the institutionalization of the precautionary principle), and more sensitive to the ecological and socio-ethical dimensions of decisions, attending in this way to societal demands for more democratic and accountable risk analysis, and for policies aligned more with public interest than with economic and industrial interests (Plaza and Todt 2005). According to the European Commission: "For Europe to become the most advanced knowledge society in the world, it is imperative that legitimate societal concerns and needs concerning science and technology development are taken on board" (European Commission 2007, p. 4).

Risk governance thus faces a fundamental tension: on the one hand, in the context of a very competitive and specialized knowledge-based economy, it aims to facilitate the controlled development of technological innovations with high economic potential; on the other, its legitimization and societal viability depends on the application, for instance, of more precautionary and democratic-participatory accountability measures, which could imply a competitive disadvantage in the techno-economic arena.

This tension is evident, for example, in the governance of the risks of nanotechnology. Nanotechnology has been called a revolution for the economy because of its capacity to operate at atomic and molecular levels and its applicability to all technological industrial sectors (Commission of the European Communities 2004, pp. 4–5). Nanotechnology is therefore a key strategic research and development area for industrialized countries and their economic competitiveness. According to the European Commission, nanotechnology R&D is an enterprise that "should not be delayed, unbalanced or left to chance" (Commission of the European Communities 2007, p. 2).[15] However, there are serious concerns about the safety of nanotechnology, since what makes nanotechnology so revolutionary and interesting — its capacity to operate at atomic and molecular sizes, transforming the way in which materials behave on larger scales regarding properties such as conductivity, lightness or resistance — also creates the potential for new risks "possibly involving quite different mechanisms of interference with the physiology of human and environmental species" (Commission of the European Communities 2008, p. 3).[16] Such concerns have prompted the EU to become the first government in the world to develop nano-specific regulations that impose special rules and conditions on the development of nano-products, mainly as a consequence of the European Parliament's legislative initiative.[17] For instance, the new regulation on cosmetic products demands a specific risk assessment for nanomaterials, mandates labeling of cosmetic products that contain nanomaterials, and requires that prior to placing a cosmetic that contains nanomaterials on the EU market the European Commission

---

[15] Market analysts foresee a world market for nanotechnology worth between €750.000 million and €2 billion and the creation of 10 million nanotechnology-related jobs — 10 % of all manufacturing jobs worldwide — by 2015. See: European Commission-Nanotechnology: "Why are nanotechnologies important for the economy, industry and job creation?" (FAQs about Nanotechology): http://ec.europa.eu/nanotechnology/faq/faqs.cfm?lg=en&pg=faq&sub=details&idfaq=28821 (Accessed on March 24, 2013).

[16] The behavior of a chemical in a nanoparticle form cannot be extrapolated from the behavior of the same material at a larger scale. The toxicity of nanomaterials is related to physical properties occurring only at molecular and atomic sizes. For example, being so small, nanoparticles are picked up by the human body and other organisms more easily than larger particles, and are able to penetrate through biological barriers inside the organisms more readily. Also, nanoparticles have a bigger surface-area-to-volume ratio than larger particles, which increases their surface energy and catalytic capacity and, in consequence, their toxicity. For thorough studies of the principles of nanotoxicology, see: Oberdörster (2010); Oberdörster et al. (2005).

[17] As the political body in the European Union that directly represents the interests of European citizens, the European Parliament is the institution within the EU showing the greatest sensitivity to consumer and environmental safety concerning nanotechnology R&D (e.g. The European Parliament 2009)

must be supplied with safety information about the product (The European Parliament and the Council of the European Union 2009; see also: Bowman et al. 2010).

However, beyond the divergences between the European Commission and the more "precautionary" European Parliament, all political bodies in the EU seem to take for granted that massive industrial development of nanotechnology is compatible with a high level of environmental and health protection, in the fundamental sense that a profound reconsideration of the economic and industrial exploitation of nanotechnology based on safety considerations would be inconceivable. For example, the European Parliament demands that the EU should invest more in the risk assessment of nanomaterials, but assumes de facto that an appropriate effort will "close the knowledge gaps" (The European Parliament 2009, p. 84; see also: Commission of the European Communities 2004). Thus it seems that nanotechnology safety (i.e., the predictability and controllability of the risks of nanotechnology) is considered a fact, perhaps because full industrial development of this technology is seen as an absolute imperative. And this despite the fact that the group of experts advising the European Commission has pointed out that nanotechnology-related risks are uncertain in a systemic way (SCENIHR 2007).

This means that more alternative, or "radical" valuations and constitutions of safety would not be considered. For instance, according to the European Commission, "Without a serious communication effort, nanotechnology innovations could face an unjust negative public reception. (...) The public trust and acceptance of nanotechnology will be crucial for its long-term development" (Commission of the European Communities 2004, p. 19). Here it is assumed de facto that any profound reconsideration of the political-industrial promotion of nanotechnology based on safety criteria is not even an option. A critical social reception of nanotechnology is conceived as "unjust" in advance, as something to be tempered by communication (i.e., dissemination) policies that would educate the ignorant publics on the real risks of nanotechnology.[18]

This attitude responds, arguably, to the tendency of regulatory institutions to frame technological appraisal according to "instrumental simple-realist questions of controlling a risk" (Wynne 2002, p. 462), i.e., according to the fundamental assumption that acceptable techno-social safety scenarios must and can be achieved on the basis of current epistemological, technological and legal principles and capacities, without renouncing basic socio-economic and political imperatives and goals. As a consequence, more radical safety scenarios are dismissed on principle. For example, in response to concerns over the health and environmental risks of nanotechnology, the Canadian non-governmental organization ETC Group (Action Group on Erosion, Technology and Concentration) called in 2002 for a "moratorium on commercial production of new nanomaterials" (ETC Group 2002, p. 6); however, the European Commission dismissed this opinion, arguing that such a measure

---

[18] In similar terms, and more recently, the European Commission, in the context of the Seventh Framework Programme for R&D, called for research on new forms of communication and social dialogue, in order to facilitate the "responsible social acceptance of nanotechnology" (European Commission 2009, 9).

would deny society the benefits of nanotechnology and could only be taken "in the event that realistic and serious risks" were identified (Commission of the European Communities 2004, p. 19). More recently, various scientific studies have drawn attention to the risks, uncertainties and extreme unknowns associated with nano-technological developments (e.g. Chandra Ray et al. 2009; Oberdörster 2010; Poland et al. 2008; SCENIHR 2007), but these have not raised profound, potentially disruptive institutional doubts about the controllability of nanotechnology (e.g. European Commission 2012).

Nevertheless, scientific capacity to assess comprehensively the risks and uncertainties of nanotechnology cannot be taken for granted. For instance, according to the European Commission, "In view of the remaining knowledge gaps, (...) potential risks of nanomaterials have to be assessed on a case-by-case basis" (Commission of the European Communities 2009, p. 8), but it has been argued that, in the light of the expected massive introduction of nanomaterials on the market, it is unrealistic to conduct a casuistic assessment of every nanomaterial: the number of nanomaterials that raise safety concerns is relatively small (mostly, nanotubes, carbon fullerenes, quantum dots, metal oxides, dendrimers and nanoscale metals), but differences in size, shape, chemistry, surface area, etc. will produce thousands of variants that will determine the environmental and health risks of nanomaterials (Walker and Bucher 2009, p. 252).

In the conviction that the industrial development of nanotechnology is an inevitable goal in the context of a hyper-competitive global knowledge economy, institutional discourses and practices concerning nanotechnology safety are probably unwilling to admit that substantive control over a world transformed at the atomic and molecular levels may not be possible (e.g. Nordmann 2005). The governance of techno-economic risk in our societies therefore represents more than an assessment and management of a set of given risks: it faces the challenge of managing a hetero-geneous variety of epistemological, technical, economic, political and cultural factors by which safety scenarios are achieved and achievable in our societies. In this sense, the political, socio-ethical and cultural considerations surrounding risks are not only relevant with regard to social acceptability, but also to the actual *constitution* of safety. For instance, according to Ortwin Renn:

"If all society would care about is to reduce the amount of physical harm done to its members, technical expertise and some form of economic balancing would suf-fice for effective risk management. However, society is not only concerned about risk minimization. People are willing to suffer harm if they feel it is justified or if it serves other goals. At the same time, they may reject even the slightest chance of being hurt if they feel the risk is imposed on them or violates their other attitudes and values. *Context* matters" (Renn 1999, p. 3050; emphasis added).

Here, societal concerns regarding techno-industrial safety are seen as "contextual," but the significance of these concerns and considerations is not just contextual but *constitutive*, in the sense that the stabilization, or normalization, of safety frameworks depends on the way in which we value and relate a heterogeneous set of epistemologi-cal, technical, economic, political and cultural factors (Rodríguez 2008, chaps. 5 and 6; see also: Healy 2004). In other words, governing the risks of progress means governing a diversity of elements that ground that progress and might need, to a greater or lesser extent, to be re-valued and re-arranged. The proposals of risk governance, even in its more "democratic," or participative, forms, may be unable to go beyond certain limits concerning potential safety scenario set-ups. For example, analytic-deliberative exercises of risk governance, in which citizens are given the chance to evaluate different technological options based on previous scientific risk assessments that determine how risky those options are in the light of various value-indicators set by societal stakeholders (e.g. Renn et al. 1993; Renn 2004), draw a clear line between scientific, socio-political and public "rationalities" (Renn et al. 1993, 193; Renn 2004, p. 330). According to this model, citizens are "value consultants" who opine about facts (Renn et al. 1993, pp. 196–197), namely about some techno-logical options (in the light of their risks) whose trajectories are not open to public scrutiny. In this respect, technological safety here is established according to certain political and economic imperatives that curtail the possibility of developing more alternative safety scenarios (Dryzek 2000, chap. 4; Dryzek et al. 2009), including alter-native risk assessments, which are not value-free (Douglas 2000; Sarewitz 2010).[19]

Safety and its legitimization therefore depends on a set of socio-economic imperatives that determine policy decisions and everyday practices. In other words, the development of alternative and arguably more legitimate safety scenarios will depend on the extent to which our societies are able to re-value and re-arrange the heterogeneous set of scientific-technological, economic, political, cultural elements that ground and inform our decisions and practices on risk. This is by no means an easy task: are our societies willing to modify the socio-economic imperatives that inform policies and everyday life? Are we willing to renounce the comforts of modern industrial societies? Are our economies willing to renounce the strategic "innovation race" and fall behind savvy competitors?

8.5 Conclusions

This paper discusses the legitimacy challenges regulatory institutions in Europe face concerning the governance and regulation of the risks of strategic technological innovations, such as agri-food biotechnology and nanotechnology. To this end, I have argued that dynamics of social mistrust toward the institutions that regulate techno-logical risks are not the mere result of the overreaction of ignorant or prejudiced citizens. The oscillation of social trust is related to a great extent to substantive epistemological, ethical and socio-political challenges of risk analysis. In the end,

---

[19] Participatory exercises like the analytic-deliberative model are not binding—i.e., policymakers are not obliged to implement, or follow, public opinions on technological options and risks. This non-binding character of citizen deliberation has been typically justified in terms of the emotional and prejudiced character of the public opinions about science and technology (e.g. Rowe and Frewer 2000, 15). As currently argued, though, this non-binding character is better understood in terms of the socio-economic pervasiveness of certain techno-industrial developments and trajectories.

risk regulations have to deal with the fundamental tension between the promotion of technological innovations (necessary in order to be competitive in a global knowledge-economy) and the strict control of health and environmental risks related to these innovations. In any case, as with agri-food biotechnology in Europe, the development of technological innovations will only be feasible if social concerns about safety are taken into account — i.e., if solid grounds for social trust in regulatory institutions are established.

However, the development and introduction of permanent risk governance measures, i.e., the implementation of more precautionary, participatory and socio-ecologically sensitive policies in risk analysis, is conditioned by a set of socio-economic imperatives, or constraints, that seriously limit the extent to which more alternative, or critical, safety scenarios for such strategic technological innovations as nanotechnology R&D can be built in our techno-industrial societies. The social legitimacy of scientific-technological progress will very likely depend upon the ability, willingness and opportunity to re-value and re-arrange the set of heterogeneous epistemological, technical, economic, political and cultural factors by which safety scenarios are built and made acceptable in our societies. Risk governance capacity to legitimize and transform will be severely curtailed unless more radical, or profound, practices and decisions are implemented with regard to safety levels in strategic techno-industrial innovations.

Acknowledgments This work was supported by the Spanish Ministry of Economy and Competitiveness (grants FFI2011-24414 and FFI2012-33550), and the Basque Government's Department of Education, Language Policy and Culture (grant IT644-13).

References

Bauer, M.W., and G. Gaskell (eds.). 2002. *Biotechnology – The making of a global controversy*. Cambridge: Cambridge University Press.

Beck, U. 1986. *Risikogesellschaft: Auf dem Weg einem andere Moderne*. Frankfurt: Suhrkampf. Translated into English by Mark Ritter: *Risk Society: Towards a New Modernity*. London: Sage. 1992.

Beck, U. 1988. Gegengifte: Die organisierte Unverantwortlichkeit. Frankfurt: Suhrkampf. Translated into English by Amosz Weisz: *Ecological Politics in an Age of Risk*. Cambridge: Polity Press. 1995.

Biegelbauer, P., and S. Borrás (eds.). 2003. *Innovation policies in Europe and the US: The new agenda*. Aldershot/Burlington: Ashgate.

Bowman, D.M., G. van Calster, and S. Friedrichs. 2010. Nanomaterials and regulation of cosmetics. *Nature Nanotechnology* 5: 92.

Butler, D. 2010, March 9. A new dawn for transgenic crops in Europe? *Nature News*. doi:10.1038/news.2010.112), www.nature.com/news/2010/100309/full/news.2010.112.html. Accessed on 27 Feb 2012.

Byrne, D. 2003. *"Irrational fears or legitimate concerns"* – *Risk perception in perspective*, SPEECH/03/593, Risk perception: Science, public debate and policy making conference, Brussels, 4 Dec 2003.

Chandra Ray, P., H. Yu, and P.P. Fu. 2009. Toxicity and environmental risks of nanomaterials: Challenges and future needs. *Journal of Environmental Science and Health. Part C, Environmental Carcinogenesis & Ecotoxicology Reviews* 27(1): 1–35.

Charnley, G., J.D. Graham, R.F. Kennedy Jr., and J. Shogren. 2000. 1998 annual meeting plenary session: Assessing and managing risks in a democratic society. *Risk Analysis* 20(3): 301–315.

Cobb, M.D., and J. Macoubrie. 2004. Public perceptions about nanotechnology: Risks, benefits and trust. *Journal of Nanoparticle Research* 6(4): 395–405.

Comité de préfiguration d'une haute autorité sur les organismes génétiquement modifiés. 2008. *Opinion on the dissemination of MON 810 on the French territory (January 9)*. Paris: French Republic.

Commission of the European Communities. 2001. *European governance: A white paper*. Brussels, 25.7.2001, COM(2001) 428 final.

Commission of the European Communities. 2004. *Towards a European strategy for nanotechnology*. Luxembourg: Office for Official Publications of the European Communities.

Commission of the European Communities. 2007. *Nanosciences and nanotechnologies: An action plan for Europe 2005–2009. First implementation report 2005–2007*. Brussels, 6.9.2007, COM(2007) 505 final.

Commission of the European Communities. 2008. *Regulatory aspects of nanomaterials*. Brussels, 17.6.2008, COM(2008) 366 final.

Commission of the European Communities. 2009. *Nanosciences and nanotechnologies: An action plan for Europe 2005–2009. Second implementation report 2007–2009*. Brussels, 29.10.2009, COM(2009)607 final.

Cranor, C. 2011. *Legally poisoned: How the law puts us at risk from toxicants*. Cambridge, MA: Harvard University Press.

Cvetkovich, G. 1999. The attribution of social trust. In *Social trust and the management of risk*, ed. G. Cvetkovich and R.E. Löfstedt, 53–61. London: Earthscan.

Cvetkovich, G., and P.L. Winter. 2007. The what, how and when of social reliance and cooperative risk management. In *Trust in cooperative risk management: Uncertainty and scepticism in the public mind*, ed. M. Siegrist, T.C. Earle, and H. Gutscher, 187–209. London/Sterling: Earthscan.

Das, T.K., and B.S. Teng. 2001. Trust, control and risk in strategic alliances: An integrated framework. *Organization Studies* 22(2): 251–283.

Dickson, D. 1984. *The new politics of science*. New York: Pantheon Books.

Douglas, M. 1985. *Risk acceptability according to the social sciences*. New York: Russell Sage.

Douglas, H. 2000. Inductive risk and values in science. *Philosophy of Science* 67(4): 559–579.

Douglas, M., and A. Wildavsky. 1982. *Risk and culture. An essay on the selection of technological and environmental dangers*. Berkeley: University of California Press.

Dryzek, J.S. 2000. *Deliberative democracy and beyond: Liberals, critics, contestations*. Oxford: Oxford University Press.

Dryzek, J.S., R.E. Goodin, A. Tucker, and B. Reber. 2009. Promethean elites encounter precautionary publics: The case of GM foods. *Science, Technology & Human Values* 34(3): 263–288.

Earle, T.C., and G. Cvetkovich. 1999. Social trust and culture in risk management. In *Social trust and the management of risk*, ed. G. Cvetkovich and R.E. Löfstedt, 9–21. London: Earthscan.

Earle, T.C., M. Siegrist, and H. Gutscher. 2007. Trust, risk perception and the TCC model of cooperation. In *Trust in cooperative risk management: Uncertainty and scepticism in the public mind*, ed. M. Siegrist, T.C. Earle, and H. Gutscher, 1–49. London/Sterling: Earthscan.

Ebert, J.D. 2012. *The age of catastrophe: Disaster and humanity in modern times*. London: McFarland and Company.

ETC Group. 2002. No small matter! Nanotech particles penetrate living cells and accumulate in animal organs. ETC *Communiqué* 76 (May/June 2002):1–8.

European Commission. 2002. *Science and society action plan*. Luxembourg: Office for Official Publications of the European Communities.

European Commission. 2007. *Work programme 2007, capacities, part 5: Science in society* (C(2007)563 of 26.02.2007). Brussels: The Seventh Framework Programme.

European Commission. 2009. *Work programme 2010, cooperation, theme 4, nanosciences, nano-technologies, materials and new production technologies – NMP* (C(2009) 5893 of 29 July 2009). Brussels: The Seventh Framework Programme.

European Commission. 2012. *Second regulatory review on nanomaterials.* Brussels, 3.10.2012, COM(2012) 572 final.

Gaskell, G. 2008. Lessons from the bio-decade: A social scientific perspective. In *What can nano-technology learn from biotechnology? Social and ethical lessons for nanoscience from the debate over agrifood biotechnology and GMOs*, ed. K. David and P.B. Thompson, 237–259. Amsterdam [et al]: Academic Press.

Giddens, A. 1990. *The consequences of modernity.* Stanford: Stanford University Press.

Giddens, A. 1994a. Living in a post-traditional society. In *Lash, reflexive modernization. Politics, tradition and aesthetics in the modern social order*, ed. U. Beck, A. Giddens, and S. Lash, 56–109. Stanford: Stanford University Press.

Giddens, A. 1994b. Risk, trust, reflexivity. In *Reflexive modernization. Politics, tradition and aesthetics in the modern social order*, ed. U. Beck, A. Giddens, and S. Lash, 184–197. Stanford: Stanford University Press.

Giddens, A. 2003. *Runaway world. How globalization is reshaping our lives.* New York: Routledge.

Giddens, A. 2009. *The politics of climate change.* Cambridge: Cambridge Polity Press.

Greenpeace. 2012. *Guía roja y verde de alimentos transgénicos* (5th ed. – Updated 8/3/2012). Madrid/ Barcelona: Greenpeace.

Greenpeace Poland. 2013, January 19. Genetically modified crops illegal – Government launches bans. In *Poland is the most recent country to ban GMO cultivation*, ed. F. Kreiss (Occupy Monsanto). www.occupymonsanto360.org/2013/01/19/poland-is-the-most-recent-country-to-ban-gmo-cultivation. Accessed on 29 March 2013.

Hansson, S.O. 2005. The epistemology of technological risk. *Techné* 9(2): 68–80.

Healy, S. 2004. A 'post-foundational' interpretation of risk as 'performance'. *Journal of Risk Research* 7(3): 277–296.

Jacob, M., and T. Hellström. 2000. Policy understanding of science, public trust and the BSE-CJD crisis. *Journal of Hazardous Materials* 78(1–3): 303–317.

Johnson, B.B. 2007. Getting out the swamp: Towards understanding sources of local officials' trust in wetlands management. In *Trust in cooperative risk management: Uncertainty and scepticism in the public mind*, ed. M. Siegrist, T.C. Earle, and H. Gutscher, 211–240. London/ Sterling: Earthscan.

Levidow, L., and J. Murphy. 2003. Reframing regulatory science: Trans-atlantic conflicts over GM crops. *Cahiers d'économie et sociologie rurales* 68(69): 47–74.

Luján, J.L., and J.A. López Cerezo. 2004. De la promoción a la regulación. El conocimiento científico en las políticas públicas de ciencia y tecnología. In *Gobernar los riesgos. Ciencia y valores en la sociedad del riesgo*, ed. J.L. Luján and J. Echeverría, 75–98. Madrid: Biblioteca Nueva/ OEI.

Marklund, G., N.S. Vonortas, and C.W. Wessner (eds.). 2009. *The innovation imperative: National innovation strategies in the global economy.* Cheltenham: Edward Elgar.

Maynard, C. 1996. Forest genetics glossary (State University of New York College of Environmental Science and Forestry). www.esf.edu/for/maynard/GENE_GLOSSERY.html. Accessed on 19 Mar 2013.

Möllering, G. 2006. *Trust: Reason, routine, reflexivity.* Amsterdam [et al]: Elsevier.

Murphy, N., and S. Krimsky. 2003. Implicit precaution, scientific inference, and indirect evidence: The basis for the US environmental protection agency's regulation of genetically modified crops. *New Genetics and Society* 22(2): 127–143.

Nordmann, A. 2005. *Noumenal* technology: Reflections on the incredible tininess of nano. *Techné* 8(3): 3–23.

Oberdörster, G. 2010. Safety assessment for nanotechnology and nanomedicine: Concepts of nanotoxicology. *Journal of Internal Medicine* 267(1): 89–105.

Oberdörster, G., A. Maynard, K. Donaldson, V. Castranova, J. Fitzpatrick, K. Ausman, J. Carter, B. Karn, W. Kreyling, D. Lai, S. Olin, N. Monteiro-Riviere, D. Warheit, and H. Yang. 2005.

Principles for characterizing the potential human health effects from exposure to nanomaterials: Elements of a screening strategy. *Particle and Fiber Toxicology* 2(8): 1–35.

Panel on Genetically Modified Organisms. 2008. Scientific opinion. Request from the European Commission related to the safeguard clause invoked by France on Maize MON810 according to Article 23 of Directive 2001/18/EC and the emergency measure according to Article 34 of Regulation (CE) No 1829/2003. *The EFSA Journal* 850: 1–45.

Pellizzoni, L. 2001. Democracy and the governance of uncertainty: The case of agricultural gene technologies. *Journal of Hazardous Materials* 86(1–3): 205–222.

Plaza, M., and O. Todt. 2005. La gobernanza de la seguridad alimentaria." Gobernanza de la ciencia y la tecnología (M. I. González and O. Todt, eds.). *Arbor* CLXXXI(715): 403–416.

Poland, C.A., R. Duffin, I. Kinloch, A. Maynard, W.A.H. Wallace, A. Seaton, V. Stone, S. Brown, W. MacNee, and K. Donaldson. 2008. Carbon nanotubes introduced into the abdominal cavity of mice show asbestos-like pathogenicity in a pilot study. *Nature Nanotechnology* 3: 423–428.

Ravetz, J. 2003. A paradoxical future for safety in the global knowledge economy. *Futures* 35(8): 811–826.

Renn, O. 1992. The social arena concept of risk debates. In *Social theories of risk*, ed. S. Krimsky and D. Golding, 179–196. Westport: Praeger.

Renn, O. 1999. A model for an analytic-deliberative process in risk management. *Environmental Science & Technology* 33(18): 3049–3055.

Renn, O. 2004. The challenge of integrating deliberation and expertise: Participation and discourse in risk management. In *Risk analysis and society: An interdisciplinary characterization of the field*, ed. T. McDaniels and M.J. Small, 289–366. Cambridge: Cambridge University Press.

Renn, O., T. Webler, H. Rakel, P. Dienel, and B. Johnson. 1993. Public participation in decision making: A three-step procedure. *Policy Sciences* 26(3): 189–214.

République Française. 2009. *Memo from the French Authorities to the Member States (Re: Draft Commission decision on the emergency measures taken by France concerning the placing on the market of genetically modified maize line MON810, pursuant to Article 34 of Regulation 1829/2003/EC, to be submitted to the vote of the Member States at the SCFCAH meeting on 16 February 2009)*. Paris: French Republic.

Rodríguez, H. 2008. *Arriskuaren eraketa dinamikak: arriskuak gobernatzeko proposamen konposizionala* (Dynamics of risk constitution: A compositional risk governance proposal). Leioa: University of the Basque Country Press. (Doctoral Thesis; in Basque.)

Rodríguez, H. 2009. La confianza pública en las instituciones reguladoras del riesgo: Tres modelos de confianza para tres desafíos del análisis del riesgo. *Argumentos de Razón Técnica* 12: 125–153.

Rowe, G., and L.J. Frewer. 2000. Public participation methods: A framework for evaluation. *Science, Technology & Human Values* 25(1): 3–29.

Sarewitz, D. 2010. Normal science and limits of knowledge: What we seek to know, what we choose not to know, what we don't bother knowing. *Social Research* 77(3): 997–1010.

SCENIHR (Scientific Committee on Emerging and Newly Identified Health Risks). 2007. *Opinion on the appropriateness of the risk assessment methodology in accordance with the technical guidance documents for new and existing substances for assessing the risks of nanomaterials*. Brussels: European Commission, Health and Consumer Protection, Directorate-General.

Shrader-Frechette, K.S. 1991. *Risk and rationality. Philosophical foundations for populist reforms.* Berkeley: University of California Press.

Shrader-Frechette, K.S. 2011. *Taking action, saving lives: Our duties to protect environmental and public health*. Oxford: Oxford University Press.

Siegrist, M., G. Cvetkovich, and C. Roth. 2000. Salient value similarity, social trust, and risk/benefit perception. *Risk Analysis* 20(3): 353–362.

Siegrist, M., H. Gutscher, and C. Keller. 2007. Trust and confidence in crisis communication: Three case studies. In *Trust in cooperative risk management: Uncertainty and scepticism in the public mind*, ed. M. Siegrist, T.C. Earle, and H. Gutscher, 267–286. London/Sterling: Earthscan.

Slovic, P. 1999. Perceived risk, trust, and democracy. In *Social trust and the management of risk*, ed. G. Cvetkovich and R.E. Löfstedt, 42–52. London: Earthscan.

The Council of the European Communities. 1990. Council Directive 90/220/EEC of 23 April 1990 on the deliberate release into the environment of genetically modified organisms. *Official Journal of the European Communities* (L 117, 8 May 1990): 15–27.

The European Parliament. 2009. European Parliament resolution of 24 April 2009 on regulatory aspects of nanomaterials (2008/2208(INI)). Nanomaterials, P6_TA(2009)0328, *Official Journal of the European Union*, C 184 E/82, 8.7.2010.

The European Parliament and the Council of the European Union. 2001. Directive 2001/18/EC of the European Parliament and of the Council of 12 March 2001 on the deliberate release into the environment of genetically modified organisms and repealing Council Directive 90/220/EEC. *Official Journal of the European Communities* (L 106/1, 17 April 2001):1–38.

The European Parliament and the Council of the European Union. 2003a. Regulation (EC) No 1829/2003 of the European Parliament and of the Council of 22 September 2003 on genetically modified food and feed. *Official Journal of the European Union* (L 268/1, 18 October 2003):1–23.

The European Parliament and the Council of the European Union. 2003b. Regulation (EC) No 1830/2003 of the European Parliament and of the Council of 22 September 2003 concerning the traceability and labelling of genetically modified organisms and the traceability of food and feed products produced from genetically modified organisms and amending Directive 2001/18/EC. *Official Journal of the European Communities* (L 268/24, 18 October 2003):24–28.

The European Parliament and the Council of the European Union. 2009. Regulation (EC) No 1223/2009 of the European Parliament and of the Council of 30 November 2009 on cosmetic products. *Official Journal of the European Union*, L 342/59, 22.12.2009.

Todt, O. 2002. *Innovación y regulación: La influencia de los actores sociales en el cambio tecnológico. El caso de la ingeniería genética agrícola*. Valencia: Universitat de València. (Doctoral Thesis.)

Todt, O. 2004. Regulating agricultural biotechnology under uncertainty. *Safety Science* 42(2): 143–158.

Walker, N.J., and J.R. Bucher. 2009. A 21st century paradigm for evaluating the health hazards of nanoscale materials? *Toxicological Sciences* 110(2): 251–254.

Winickoff, D., S. Jasanoff, L. Busch, R. Grove-White, and B. Wynne. 2005. Adjudicating the GM food wars: Science, risk, and democracy in world trade law. *The Yale Journal of International Law* 30(1): 81–123.

Wynne, B. 1992. Uncertainty and environmental learning. Reconceiving science and policy in the preventive paradigm. *Global Environmental Change* 2(2): 111–127.

Wynne, B. 1996a. Misunderstood misunderstandings: Social identities and public uptake of science. In *Misunderstanding science? The public reconstruction of science and technology*, ed. A. Irwin and B. Wynne, 19–46. Cambridge: Cambridge University Press.

Wynne, B. 1996b. May the sheep safely graze? A reflexive view of the expert-lay knowledge divide. In *Risk, environment & modernity*, ed. S. Lash, B. Szerszynski, and B. Wynne, 44–83. London [et al]: Sage.

Wynne, B. 2002. Risk and environment as legitimatory discourses of technology: Reflexivity inside out? *Current Sociology* 50(3): 459–477.

Wynne, B. 2006. Public engagement as a means of restoring public trust in science – Hitting the notes, but missing the music? *Community Genetics* 9(3): 211–220.

Chapter 9
The Social Dimension of Technology: The Control of Chemical and Biological Weapons

Brian Balmer

9.1 Introduction

Very shortly after the September 11 2001 terrorist attacks on the twin towers in New York and the Pentagon building in Arlington, Virginia, some time around September 18th, the first of two batches of letters was posted in the Princeton, New Jersey area to various media outlets, one in Florida the rest in New York. The letters warned that they contained anthrax and, although initially dismissed as hoaxes, they had indeed been laced with deadly anthrax spores (Cole 2003; Guillemin 2011). On October 3rd the first victim was diagnosed and died a week later. As others were diagnosed, the letters were confirmed as the source of the anthrax infections. Then, a further set of letters was mailed at some time around October 8th, including two to Democrat senators in Washington DC. Eventually, there were 12 deaths from anthrax. In their ensuing investigation of the attacks, the Federal Bureau of Investigation's (FBI) lead theory was that they were perpetrated by a domestic criminal, perhaps a disgruntled employee of the US biological warfare (BW) defence programme. Some time later, the FBI named Bruce Ivins, a senior scientist at the Army Medical Research Institute of Infectious Diseases, Fort Detrick, as their prime suspect. Ivins committed suicide in July 2008, so the FBI was not able to bring a case to court.

Some years before the anthrax attacks a Japanese sect called Aum Shinrikyo launched terrorist attacks, not with living, disease-causing biological weapons, but with highly toxic chemical weapons (Tucker 2007). The first attack in Matsumoto, a tourist and industrial city some 100 miles northwest of Tokyo, took place on 27th

Brian Balmer is supported by AHRC grant AH/K003469/1

B. Balmer (✉)
Department of Science and Technology Studies, University College London,
Gower Street, London WC1E 6BT, UK
e-mail: b.balmer@ucl.ac.uk

© Springer International Publishing Switzerland 2015   167
W.J. Gonzalez (ed.), *New Perspectives on Technology, Values,
and Ethics*, Boston Studies in the Philosophy and History of Science 315,
DOI 10.1007/978-3-319-21870-0_9

June 1994. The cult members used a truck designed to disperse sarin gas (sarin is a chemical agent which rapidly attacks the nervous system). They parked outside a supermarket and released the sarin, which dispersed to nearby apartments and killed 7 people and injured around 250. This was dismissed by Japanese authorities as a chemical hobbyist's accident, a tragic event after someone had been playing around with pesticides. The second attack took place on 20th March 1995, and this time the cult targeted the Tokyo underground system. Here, the sarin was simply placed in sealed plastic bags, which the cultists then pierced before swiftly getting away from the deadly bags. Then the sarin gradually dispersed, causing mayhem and killing around 12 people, leaving 17 in a critical condition, a further 37 with serious symptoms (shortness of breath, vomiting, severe headache, muscular twitching, gastrointestinal problems) and around a 1000 others with mild symptoms.

Both attacks highlight the on-going reality of the threat from chemical and biological weapons. Added to these examples, it is sobering to remember that the major interest shown in these weapons in the twentieth century was by nation states, with many launching highly secretive, state-sponsored, efforts to apply science and technology to the creation of ever more deadly chemical and biological agents (Wheelis et al. 2006; Coleman 2005; Guillemin 2005; Harris and Paxman 1982). However, we should also not lose sight of the many efforts there have been to control the threat from these agents and this more hopeful topic will be the main focus of the present chapter. It is written for an audience unfamiliar with this topic and is intended to point readers to some key issues that could potentially benefit from scrutiny by historians, philosophers and sociologists of science. The first part of this chapter will simply outline the key aspects of the international control of chemical and biological weapons, focusing on the main treaties that outlaw these weapons. It is necessarily descriptive. I move from this description, to highlighting some of the key practical and conceptual dilemmas in implementing these controls. These are the problems of definition, the problem of 'dual-use', the problem of distinguishing defensive from offensive research, the problem of verification, and the difficulty of analyzing chemical and biological warfare from a cultural perspective. Also in this second part, I will discuss some of my own historical research on chemical and biological warfare and add some brief reflections on how this work might inform some of the contemporary practical issues in the control of chemical and biological warfare.

9.1.1 Definitions

The definitions of chemical and biological weapons are not always clear-cut, as will be discussed later in the chapter. That said, as a way into this topic, it is worth providing some general, conventional definitions. Biological weapons—often dubbed as germ weapons—deliberately employ living organisms (usually bacteria, viruses or fungi) to cause incapacitation or death (Dando 1994). Chemical weapons are non-living and generally regarded as a class of weapons where toxicity—rather than other means such as explosive blast—is the main cause of death. They have generally been categorized as acting through blistering, choking, poisoning the blood, and interfering with the nervous system to incapacitate or kill (Kenyon 2000). Intermediate classes of weapons exist, such as toxins, which are non-living chemicals but derived from living organisms. In all cases, the targets of these weapons can be humans, non-human animals, or plants (such as agricultural crops).[1]

9.1.2   Chemical and Biological Arms Control

The first large-scale use of chemical weapons was during the First World War, although there are many examples of more limited use of chemical and biological weapons dating back to antiquity (Mayor 2003). There had been attempts to limit potential chemical weapons use before the First World War, but following the use of chemicals such as chlorine, phosgene and mustard gas in that war, fresh attempts were made during the inter-war years to ban chemical weapons. These efforts focused on negotiations by the League of Nations, and resulted in the 1925 Geneva Protocol. This was a treaty which outlawed both chemical ("use in war of asphyxiating, poisonous or other gases, and of all analogous liquids, materials or devices") and "bacteriological" warfare. The treaty had shortcomings. What counted as "use in war" remained unspecified. Some nations, notably the USA, did not ratify the treaty, while many others tabled reservations such that the Protocol was effectively an agreement to "no-first-use" of these weapons rather than an outright ban.

From the end of the 1950s chemical and biological disarmament came back into the foreground of international discussion. UK and USSR proposals for programmes of disarmament that would encompass both chemical and biological warfare were put to the United Nations (UN) General Assembly in September 1959 and passed to the Ten-Nation Disarmament Committee for detailed discussion (SIPRI 1971, Vol IV). Their deliberations eventually led to a joint US-Soviet draft treaty on general and complete disarmament covering nuclear, chemical, biological and other weapons of mass destruction. Discussion of some form of chemical and biological weapons (CBW) control or ban also took place within the UK and USA in 1963, although this was largely dismissed as it might provide the USSR with a reason to pursue options for nuclear disarmament (Chevrier 2006).

Success in international nuclear arms limitation negotiations, namely the Limited Test Ban Treaty (1963) and later the Nuclear Non-Proliferation Treaty (NPT) (1968), kept arms control and disarmament initiatives moving. The Vietnam War, however, threw a shadow over the progress of the chemical and biological warfare treaty negotiations. More so, because the USA had made use of tear gas in Vietnam, and was one of the only major powers to have signed, but not ratified, the 1925 Geneva Protocol. Existing literature (Wright 2002; Chevrier 2006; SIPRI 1971;

---

[1] Some chemical and biological agents are particularly persistent in the environment, so also could in principle be used to contaminate land, equipment or buildings.

Sims 1988, 2001; Walker 2012) takes us through key events: a Hungarian proposal to the UN singled out the USA, although it called for all states to observe the provisions of the Geneva Protocol; a UK policy review and Draft Biological Weapons Convention was produced in July 1969; a Soviet draft convention in September 1969 proposed prohibitions on development and production of both chemical and biological weapons; this was followed by announcements by President Richard Nixon that the USA would unilaterally disarm its biological and then toxin weapons (Tucker and Mahan 2009); a further twist in the negotiations was a Soviet draft proposal conceding that they would accept separate chemical and biological weapons conventions.

The Biological Weapons Convention (BWC) was opened for signature on 10 April 1972 and entered into force on 26 March 1975.[2] The treaty currently has 170 state signatories, with several nations joining as recently (at the time of writing) as early 2013. It stands as the first international treaty to outlaw an entire class of weapons. And, it is also a remarkably short treaty that is but a few pages in length. The heart of the treaty is Article 1 where each state party that has signed the treaty:

"…undertakes never in any circumstances to develop, produce, stockpile or otherwise acquire or retain:

1. Microbial or other biological agents, or toxins, whatever their origin or method of production, of types or in quantities that have no justification for prophylactic, protective or other peaceful purposes:

2. Weapons, equipment or means of delivery designed to use such agents or toxins for hostile purposes or in armed conflict" (United Nations 1972).

Although negotiation for the Chemical Weapons Convention (CWC) began alongside the BWC negotiations in August 1968, the CWC was agreed and then opened for signature much later, in January 1993. Negotiations went through many shifts in the geo-political climate, including the Vietnam War, Cold War détente and re-escalation, then its demise and then the first Gulf War. This meant protracted negotiations in changing geo-political circumstances, but the treaty eventually entered into force 180 days after ratification by the 65th signatory country (Hungary) and entered into force on 29 April 1997.

The treaty is a much longer document than the BWC, (see www.opcw.org), with 24 articles and 3 annexes. Its main provisions are that:

• Each state party undertakes never, in any circumstances, to: acquire, stockpile or retain chemical weapons or transfer, directly or indirectly, chemical weapons to anyone.

• Use chemical weapons, or engage in military preparations for doing so.

• Assist, encourage or induce, in any way, anyone to engage in any activity prohibited by the treaty.

• Disarmament: a commitment to destroy, within 10 years, a signatories' own chemical weapons, production facilities, and weapons abandoned on the territory of another State Party.

---

[2]When a treaty enters into force its provisions become legally binding on states parties to the convention.

Arms Control treaties remain at the heart of chemical and biological weapons control but are not only way of controlling these weapons. Total disarmament, coercion and sanctions are examples of other means. Dando (2002) points out that any arms control treaty will:

• Establish a "regime" for governing a weapon technology. A regime consists of norms, laws, organisations etc. In other words, the treaty is not just a piece of paper;

• Be written so that the benefits of signing the treaty should outweigh the costs;

• Need always to balance a range of interests, particularly as chemical and biological warfare have strong overlaps with civil research in areas such as agricultural and pharmaceutical research. In this respect, treaties such as the BWC and CWC are not just about international security—they affect industry and the scientific community.

Moreover, Dando also points out that there are different functions of verification. An ideal treaty would catch violators, but can also encompass more modest aims, such as setting up a framework for expressing concerns, providing hurdles for would-be perpetrators and suchlike.

So, as mentioned, while treaties are a crucial means to prevent the misuse of chemical and biological weapons, they are only one aspect of what has been called the "web of deterrence" (BMA 1999). Other ways of controlling these weapons include export monitoring and control, research on defensive and protective measures, national and international responses to chemical and biological acquisition and/or use, and the introduction of codes of conduct for scientists. Rather than describing these other measures in any detail, I want to move to a discussion of the problems of controlling chemical and biological weapons. This, I think, raises some more conceptual issues which I believe scholars of history, philosophy and social studies of science are equipped to address.

9.2   Problems of Controlling Chemical and Biological Weapons

9.2.1   Definitions

What counts as a biological agent? How toxic does a chemical have to be to count as a chemical agent? While standard definitions were offered at the start of this chapter, these still raise problems for controlling weapons. So, it is notable that the two treaties do not give a list of banned substances. This is for a variety of reasons. As will be discussed, many chemical and biological agents are "dual-use," in other words they can be used for benign and malign purposes. A list would also leave the treaty open to the weakness that future novel agents, such as newly synthesized chemicals or genetically modified organisms, could be construed as being outside the scope of the treaties.

The treaties both circumvent these problems by employing the so-called General Purpose Criterion. This does not regulate specific agents, but rather it regulates intent. So, for instance, the BWC states that it bans: "Microbial or other biological agents, or toxins, whatever their origin or method of production, of types or in quantities that have *no justification for prophylactic, protective or other peaceful purposes*" (emphasis added). Possession of the same biological agent could be justified if, for example, it was demonstrated that it was being used to make vaccines but would be banned if it was proved that the agent was intended for a bomb.[3] While the General Purpose Criterion avoids the problems of making lists, it also has the disadvantage of making compliance and verification very difficult, as there is now a need to prove intent.

9.2.2   The Problem of Dual-Use

As mentioned, a central problem for implementing controls over chemical and biological warfare is the so-called "dual-use" dilemma. Buchanan and Kelley (2013) argue that there are actually two ways in which the term 'dual-use dilemma' is used. First, the term is applied to benign research that is appropriated by non-state terrorists or by another state with aggressive intentions. Second, it is applied to problems that arise when benign research is appropriated by a country's own government to develop weapons.

The history of both biological and chemical weapons lies in civil research—health research and the chemical industry—and it is possible to envisage peaceful and harmful uses for many agents. So, for example, the chemical thiodiglycol is a key precursor for mustard gas, but is also used to produce ink for ballpoint pens. Another example is the toxin ricin, which is extracted from the beans of the castor plant; it is cultivated commercially for castor oil, which beyond medicinal uses has many industrial applications (e.g., in manufacture of paint resins, varnishes, nylon-type synthetic polymers, cosmetics and insecticides) (Tucker 1994). It is also highly toxic and has been investigated as a potential weapon in the USA, Canadian, British, French and Japanese biological weapons research programmes during the twentieth century.

Dual-use means that problems that arise are not confined to research at military establishments. One concern that has been much debated since the 2001 anthrax attacks is when scientists in civil settings undertake research for peaceful purposes, but might serendipitously discover something with potentially malign use.[4] In 2001,

---

[3] The CWC also lists 14 families of chemicals and 29 individual chemicals listed for control and grouped into three 'schedules' (1 being most toxic), with obligations to declare, restrictions on amounts possessed and produced per annum. Despite this list, states still have obligations to control all chemicals under the General Purpose Criterion, not just those listed.

[4] Although civilian science became a focus of concern following the 2001 anthrax attacks, the predominant theory at the time was that the perpetrator was someone from within a military research establishment.

for example, a potential problem arose from the research of scientists in Australia working on mouse pox virus with the aim of controlling the pest in domestic and agricultural settings. The genetically modified virus was meant to act by "tricking" the mouse immune system into reacting against its own eggs, effectively sterilizing the mouse. In fact, scientists accidentally created a highly lethal version of the virus, which wiped out their experimental mouse population. In turn, this raised issues of whether or not this new knowledge could be applied, for instance, to the related smallpox virus (Selgelid 2007). More recently, at the end of 2011 and into 2012, there was much debate in scientific and policy circles about whether or not new research on genetically modified H5N1 avian influenza virus should be published in leading scholarly journals. The work had produced a strain of the virus that could readily transmit between ferrets. With echoes of the mousepox debate, concerns were expressed about the implications of this work for biosecurity, and whether or not the entire paper or the methods section should be redacted.[5]

When considering cases such as this and whether or not scientific work should be published it is important, first, to think about whether any knowledge is fully open or secret or, instead, whether there is a spectrum of disclosure. Journal publications are only one aspect of how science gets communicated, so a focus on them as the main place where new science becomes public may not capture the full picture of how science gets done (Rappert and Balmer 2007). A different point, in relation to the same debates about serendipitous discovery and dual-use, is that dual-use is often framed as balancing just two values: scientific openness and national security (Buchanan and Kelley 2013). Buchanan and Kelley (2013) instead argue that multiple values are at stake in these debates including the value of producing knowledge; the protection of human and animal subjects in experiments; advancing public health protection; academic autonomy; and the constraint of government power (governments are key players in deciding whether or not an emergency is sufficiently grave for security considerations to outrank all the other values at stake).

9.2.3   A Case Study of Dual-Use: The Discovery of the V-Series Nerve Gases

Underlying the issues raised in the previous section is a further conceptual dilemma about what makes a substance dangerous, and whether "dual-use" is an inherent property of any particular agent. I want to turn now to a more detailed case study from my own research, undertaken in collaboration with Dr Caitriona McLeish at

---

[5] It is not difficult to find a range of divergent views on this topic. A good starting point is a short video made by *Nature*: http://blogs.nature.com/news/2012/02/video-debating-h5n1-and-dual-use-research.html (accessed on 23/04/13). For a philosophical treatment of the issues raised see Evans (2013).

the University of Sussex in the UK, to illustrate how historical, social and philo-sophical studies of science and technology might provide some insight into this more conceptual issue about dual-use. The case study is a summary of our recent archival work on the history of the nerve gases in the UK during the Cold War (McLeish and Balmer 2012).

The nerve gases are highly toxic organophosphorus compounds (organic com-pounds containing phosphorus). The G and V-agents are each a subset of this class of chemical warfare agents, designated as nerve gases because they interfere with the transmission of nerve signals across the junctions (synapses) between nerve cells (neurons) or between a neuron and an organ that is activated by a nerve impulse (an effector organ) such as a muscle. Nerve agent interferes with the chemical signal (acetylcholine) by blocking the enzyme that breaks this chemical down once it has performed its signalling function (cholinesterase). In effect, the nerve continues to signal and signal without switching off. Exposure to lethal nerve agents results, within minutes, in major convulsions, then as the breathing muscles stop function-ing, flaccid paralysis ensues and culminates in death by asphyxiation.

Historically, the nerve gases passed through two generations. The first genera-tion, known as the G agents, were discovered in the course of pesticide research a few years prior to the outbreak of the Second World War, initially by Gerhard Schrader, a scientist working for the chemical giant, IG Farben (Tucker 2007). Schrader worked on fumigants that would target weevils in grain and fleas in ships and domestic settings. Schrader tried synthesizing and testing different compounds, eventually using organophosphorus compounds which proved to be highly toxic to humans, thus making them unsuitable as pesticides. Nonetheless, this very property meant they were potential weapons. A 1935 Reich official ordinance required all new discoveries and patents of potential military significance to be reported to the War Office (which could declare them secret). IG Farben passed results of the new compound to the War Office, who in turn passed the information to the German Army's Gas Protection Laboratory in Spandau, Berlin. Here it was developed as the first of a series of G agents (named tabun), and passed back to IG Farben for manu-facture, although eventually neither side used chemical weapons in the Second World War.

The G agents were discovered by the Allies when they found German munitions filled with agent Tabun in the closing stages of the war (Tucker 2007; Harris and Paxman 1982). Work on the G agents soon commenced at the UK's Chemical Defence Experimental Station (in 1948 renamed the Chemical Defence Experimental Establishment [CDEE]) at Porton Down in Wiltshire. But, although this research commenced in a post-war policy environment where research on chemical and bio-logical weapons was ranked highly in defence policy and therefore encouraged (Balmer 2001), by the early 1950s there were suggestions from the CDEE's scien-tific advisors that the work on the G agents had reached its limits.

To help provide new avenues for research, CDEE scientists approached the chemical industry via its parent ministry, the Ministry of Supply. Staff at the Ministry, in turn, wrote to the trade association for the chemical industry, the Association of British Chemical Manufacturers (ABCM), in 1951 asking them to contact their member firms for information on any newly discovered, highly toxic compounds. This approach met with little success as firms expressed concerns about protecting their commercial secrets. A fresh round of letters was issued in 1953, this time with the promise that anything passed to the Ministry would end up at CDEE with simply a code letter (C) and number. Once again, this effort did not provoke much response but did encourage the firm Imperial Chemical Industries (ICI) to allow the Ministry to take one of its newly discovered organophosphorus pesticides, Amiton, "well within the barbed wire fence".[6] In the context of dual-use, it is interesting to note that a form of Amiton was later launched onto the market as an insecticide in 1957, but it was not a successful insecticide because it was too toxic for use *by* humans. As with Tabun in the first generation of nerve gas research, this characteristic meant it was potentially toxic enough for use *against* humans.

By November 1953 both C11 (the code name assigned to Amiton) and another related compound were acknowledged by scientists to have provided "an entirely new lead in the nerve gas field".[7] Over the course of the next few years, the new information was sent to collaborative counterparts in the USA and Canada. Many related compounds were synthesized in this new V-series (the V apparently standing for venomous), and in February 1957 the US Army Research and Development Command selected agent VX as the most promising. Shortly afterwards, the Ministry of Supply once again tried their letter-writing tactic, alongside visits to firms. This time, several firms responded negatively to the call for their results. Their reasons for being unable to oblige were expressed along the lines that, while the firms might occasionally discover compounds highly toxic to humans, as pesticide firms this would usually be a signal to change direction. In other words, they had little prior incentive to have pursued work to a stage that could be useful to the CDEE scientists.

This final point underlines a more general conceptual point about dual-use that emerges from this case study. Put bluntly, Amiton was not simply dual-use but had to be made dual-use. A narrow focus on the technology alone gives rise to the dual-use "dilemma," where the malign and the benign are construed as inherent in the technological artefact (McLeish 2007). On the other hand the military's efforts to capture and transform a pesticide into a chemical weapon involved active configuration of a network of artifacts and people: letters had to be carefully worded and written; secrets had to be negotiated and worked around; the different goals of the chemical industry and military had to be brought into alignment. Readers familiar with actor-network theory would refer to this configuration process as a form of "heterogeneous engineering" to build a successful network (Latour 1987). In short, this case study suggests that technology transfer, rather than being a simple hand-

---

[6]TNA, WO 188/2721. Letter J McCaulay to RM Winter, Research Controller, Messrs ICI Ltd, Nobel House, Buckingham Gate (15 July 1953). Amiton was actually discovered by scientists working at Plant Protection Ltd, a subsidiary firm of ICI (see Mcleish and Balmer 2012).

[7]TNA, WO195/12549. Ministry of Supply. Chemical Defence Advisory Board. Minutes 24th Meeting of the Board (5 November 1953).

over of the "same" knowledge, involved the forging and coordination of all manner of invisible infrastructure to turn the pesticide into a weapon.

9.2.4   The Problem of Defence/Offence Overlap

There is also a blurred distinction between offensive and defensive biological and chemical warfare (Strauss and King 1986). Countries that have signed up to the BWC and CWC have committed to abandon offensive work on CBW, but are permitted to continue research on defensive measures. Now, while improving a gas mask, for example, is clearly a defensive measure, there are examples of far more ambiguous "defensive" measures.[8] In 2001, three US journalists uncovered highly secretive projects undertaken in the USA: to build a biological bomb based on a Soviet blue-print; to build a biological weapons factory in the Nevada desert using only material bought from legal sources such as hardware stores; and research on a vaccine-resistant strain of anthrax (Miller et al. 2001). When these events were made public, they were justified by the authorities as defensive research—but, at the same time, they raise the question of just where the line would be drawn and how the USA would have reacted to other countries undertaking the same research.

9.2.5   The Problem of Verification

How would we know if someone had violated the treaties and used chemical or biological weapons? Investigating alleged CBW attacks is not straightforward, particularly with biological weapons where it may be difficult or impossible to differentiate between a natural or deliberate outbreak of a disease (Clunan et al. 2008). One such example occurred in the former Soviet Union in 1979. In the early 1980s reports began to appear in the west about an unusually large outbreak of anthrax that started in April 1979, at Sverdlovsk, some 900 miles east of Moscow (Gordin 1997; Guillemin 1999; Meselson et al. 1994; Leitenberg and Zilinskas 2012). In a city of 1.2 million people, hundreds of residents were reported to have died over a 1 month period.[9] The USA raised the issue at the 1980 international review meeting of the BWC, demanding an explanation.[10] The Soviets claimed that the outbreak had been caused by people eating contaminated meat which had been sold on the black market. The alternative explanation was that anthrax had been released following an

---

[8] Although a sceptic would point out that even a gasmask would be needed to protect an aggressor using chemical weapons from being affected by their own weapon, so even this item is not so straightforwardly defensive.

[9] Later investigations put the number lower (Guillemin 1999).

[10] The review meetings take place every 5 years in Geneva and are an opportunity to review the development of the BWC regime. Similar provisions are made in the CWC.

accident at a nearby military facility. The USA did not make a formal charge against USSR for violating article 1 of the Biological Weapons Convention; instead they preferred to tackle the matter through a less politically sensitive article of the convention, article 5, whereby states parties agree to consult bilaterally and multilaterally to solve any problems with the implementation of the BWC. This resulted in a series of diplomatic démarches to the Soviets, first by US and then UK ambassadors, each producing the same reiteration of the contaminated meat explanation. Moreover, in the charged atmosphere of the collapse of détente (the Soviet invasion of Afghanistan had taken place only months before), the Soviets complained that the outbreak at Sverdlovsk had nothing to do with the BWC and that the accusations were simply a Western ploy to divert attention away from substantive arms control issues.

Documents just emerging from the UK archives reinforce existing accounts about just how difficult it was at the time to establish definitively the cause of the outbreak, even though the US intelligence community knew that a secret military facility, believed to be a biological weapons research facility, was sited in Sverdlovsk.[11] Explanation of the outbreak was underdetermined in two significant and distinct ways: causation and legality. According to the UK Foreign and Commonwealth Office, the "US were confident that there had been a serious outbreak of human anthrax at Sverdlovsk at the beginning of April 1979. They were not confident of the cause, nor whether it indicated a violation of the BWC."[12] There was conflicting and incomplete data concerning, among other things, the pattern and length of the outbreak, the number of deaths, and the quantity of anthrax involved. Moreover, even assuming that there had been an accident, it could have involved defensive research that was permitted under the treaty. So, even if the outbreak had been traced back to the facility, there would have been no way to establish if the activity taking place there was defensive or offensive research. All of these problems underline the difficulties of verifying both the biological and chemical weapons conventions.

9.3   The Culture of Chemical and Biological Warfare

The final section of this chapter is less about a problem faced in controlling chemical and biological weapons, and more about attempts to try and illuminate arms control discussions by thinking about cultural aspects of these weapons: how are our attitudes towards chemical and biological warfare embedded in either deeply held beliefs, or conversely, in more routine beliefs and practices that may be "so obvious" they often escape our attention? In this vein, Price (1997) and Jefferson (2009) have wrestled with the seemingly nebulous idea of, respectively, a chemical

---

[11] TNA, FCO 66/1520 and FCO 66/1521.

[12] TNA FCO 28/4205 Draft attached to A, Reeve (Arms Control and Disarmament Department) (15 September 1980).

or biological weapons taboo. This is the deep-seated horror and revulsion that is commonly provoked by these weapons. To understand the chemical weapons taboo, Price argues that it is important to look beyond any intrinsic features of these weapons and instead attend to the changing context in which the taboo has been negotiated. Price is not suggesting that these weapons have no intrinsic characteristics, but finds the appeal of this essentialist approach to their ontology lacking. At the same time he rejects the idea that only useless weapons get banned. In this vein he contends that:

"Neither the view that chemical weapons are of no military utility nor the assumption that the taboo is simply explained by the unique physical characteristics of the weapons suffices to provide a fully satisfactory account of the resilience of the taboo" (Price 1997, ix).

His detailed historical analysis reveals the meaning of the taboo in flux: changing and contested, but always bound up with far wider issues, such as struggles to define what it means to be a civilized nation. While this anti-essentialist stance, with its emphasis on historical contingency, might seem unsettling as it implies the roots of the taboo may be deep but are not ahistorical, it is also a reminder that moral revulsion towards chemical (and biological) weapons is something that cannot be taken for granted and has to be (and, more optimistically, can be) constantly made and re-made through our own efforts.

Shifting from taboo and the idea of culture as deep-seated belief, several scholars have turned their attention to culture construed as the routine and taken-for-granted. This work examines the day-to-day culture of laboratory life and, with respect to arms control, brings a sociological dimension to philosopher Michael Polanyi's notion of "tacit knowledge" (Polanyi 1958). The idea that some knowledge can be codified, but some knowledge is "tacit," in other words difficult or impossible to codify (so-called "riding a bike knowledge"), was first borrowed and used in sociology of science by Collins (1985). He wanted to explain why, in his ethnographic work with physicists, the seemingly straightforward act of replicating a laser built by one group of scientists, was actually extremely difficult for other groups. Collins found that face-to-face contact and extended time spent in different laboratories proved to be important for moving beyond written instructions and acquiring the tacit knowledge needed to make the laser work.

This analysis in terms of tacit knowledge and skills has been imported into an analysis of nuclear arms control by MacKenzie and Spinardi (1995), and more recently applied by Vogel (2008) to the area of biological weapons control. Vogel sets out to challenge the idea that it is sufficient to publish scientific results in order for another (in this case, terrorist) group of scientists to replicate those results. This is one concern voiced in both the mousepox and H5N1 cases discussed earlier in this chapter. In contrast, Vogel points out that in addition to the written paper, any putative replication of the results would require laboratory skills, scientific expertise, prior training and infrastructure. Moreover, in her interviews with two groups of scientists, one that had artificially synthesised poliovirus and another that had artificially synthesised phiX bacteriophage virus, Vogel's interviewees constantly described the many points in their daily research processes that relied on tacit knowledge. Her findings are not a message to become complacent about possible malign uses of the new life sciences, but at the same time they challenge what she calls the "dominant framing" of the bioterrorist threat, which portrays scientific knowledge as straightforwardly transferrable.

My last point in this section is a suggestion, rather than assertion, that the relationship between gender and CBW might prove to be interesting and illuminating. There is a small but significant literature on how gender shapes the design of military technologies (e.g. Cohn 1987; Edwards 1990; Weber 1997). Here, I provide two hints that this literature could inform the study of chemical and biological weapons; both hints concern a recurrent historical preoccupation with women and poison. Historian, Adrienne Mayor (2003) refers to this link between women and poison when she discusses the ancient motif of the poisonous woman—or the poison maiden—mentioned as early as AD 1050 in Sanskrit literature. These legends refer to dancing maidens that were sent as a gift to the enemy, but once accepted into the enemy camp, Mayor writes, "a touch, a kiss, or sexual intercourse with one of these ravishing but deadly damsels brought death" (p. 142).

Kord (2009) supplies a second study of the links between gender and poison. She challenges the claim that there is a natural (biological or psychological) link between weakness, women and poison. As her starting point, she cites an overt attempt to naturalise this link, from Hans Gross' *Criminal Psychology* (1898):

"It is well known that poison murder is predominantly committed by women... All kinds of murder require courage, willpower and physical strength, poison murder alone does not necessitate any of these characteristics, and since women possess none of them, they automatically murder by poison. There is nothing strange or remarkable about this, it follows logically from female characteristics familiar to us all. Thus it makes sense, when in doubt regarding a murder by poison, to suspect a woman in the first instance and a weakly, effeminate man in the second" (quoted in Kord 2009, pp. 154–155).

Kord points out that any statistical correlation between gender and the means of murder is simply false. She also notes that the link assumes a crude biological determinism. Historically, it illustrates a common theme in feminist critiques of science, that a supposedly neutral science (psychology) can draw on common and unspoken assumptions about gender identity (Schiebinger 2001). In the context of this chapter, it is interesting to note that these associations are simultaneously depictions of both the wielder of the weapon and the nature of the weapon: to speak of one, is to speak of the other. I cannot develop these ideas any further in this chapter, except to note that both Mayor and Kord might provide a doorway for thinking more deeply about how gender and, in this instance, chemical warfare intertwine.

9.4 Conclusions

This chapter has provided an overview of some of the problems facing the control of chemical and biological weapons. I have tried to demonstrate how historians, philosophers and sociologists of science might bring different perspectives to the debate by challenging widely held assumptions and providing conceptual clarity. Because this chapter is written for a collection aimed primarily at philosophers of science, I want to conclude by drawing out three areas where I (as someone who is not a trained philosopher of science) think these issues could benefit from the philosopher's insights. The most obvious area is in the ethics of science and technology, and professional bioethicists, some cited in this chapter, have already turned their attention to the H5N1 and mousepox cases described in this chapter, and to the dual-use problem more generally. The second area relates to ontology. The nerve agent case study has, at its heart, a problem about what chemical substances are, and whether we understand this in an essentialist or more relational way. Finally, when considering verification, the problem of underdetermination—much debated within philosophy of science—looms large. Underdetermination is not just a philosophical puzzle confined to the laboratory, but as the Sverdlovsk incident demonstrates, it is a practical dilemma facing the politically charged area of arms control. Besides ethics, philosophy of science may seem distant from the problems of chemical and biological warfare but this chapter has made a preliminary attempt to draw these two worlds together.

References

Balmer, B. 2001. *Britain and biological warfare: Expert advice and science policy, 1935–1965*. Basingstoke: Palgrave.

BMA. 1999. *Biotechnology, weapons and humanity*. London: British Medical Association.

Buchanan, A., and M. Kelley. 2013. Biodefence and the production of knowledge: Rethinking the problem. *Journal of Medical Ethics* 39: 195–204.

Chevrier, M. 2006. The politics of biological disarmament. In *Deadly cultures: Biological weapons since 1945*, ed. M. Wheelis, L. Rózsa, and M. Dando. Cambridge, MA: Harvard University Press.

Clunan, A., P. Lavoy, and S. Martin. 2008. *Terrorism, war, or disease?: Unraveling the use of biological weapons*, Stanford security studies. Stanford: Stanford University Press.

Cole, L. 2003. *The anthrax letters: A medical detective story*. Washington, DC: Joseph Henry Press.

Cohn, C. 1987. Sex and death in the rational world of defense intellectuals. *Signs* 124: 687–718.

Coleman, K. 2005. *A history of chemical warfare*. Basingstoke: Palgrave.

Collins, H. 1985. *Changing order: Replication and induction in scientific practice*. Chicago: University of Chicago Press.

Dando, M. 1994. *Biological warfare in the 21st century*. London: Brasseys.

Dando, M. 2002. *Preventing biological warfare: The failure of american leadership*, Global issues. Basingstoke: Palgrave MacMillan.

Edwards, P. 1990. The army and the microworld: Computers and the politics of gender identity. *Signs* 16(1): 102–127.

Evans, N. 2013. Great expectations – Ethics, avian flu and the value of progress. *Journal of Medical Ethics* 39: 209–213.

Gordin, M. 1997. The anthrax solution: The Sverdlovsk incident and the resolution of a biological weapons controversy. *Journal of the History of Biology* 30(3): 441–480.

Guillemin, J. 1999. *Anthrax: The investigation of a deadly outbreak*. Berkeley: University of California Press.

Guillemin, J. 2005. *Biological weapons: From the invention of state sponsored programs to contemporary bioterrorism*. New York: Columbia University Press.

Guillemin, J. 2011. *American anthrax: Fear, crime and the investigation of the nation's deadliest bioterror attack*. New York: Henry Holt.

Harris, R., and J. Paxman. 1982. *A higher form of killing: The secret story of gas and germ warfare*. London: Chatto and Windus.

Jefferson, C. 2009. *The taboo of chemical and biological weapons: Nature, norms and international law*. Unpublished DPhil dissertation, SPRU, University of Sussex.

Kenyon, I. 2000. Chemical weapons in the twentieth century: Their use and their control. *The CBW Conventions Bulletin* 48: 1–15.

Kord, S. 2009. *Murderesses in German writing: 1720–1860: Heroines of horror*. Cambridge: Cambridge University Press.

Latour, B. 1987. *Science in action: How to follow scientists and engineers through society*. Cambridge, MA: Harvard University Press.

Leitenberg, M., and R. Zilinskas. 2012. *The soviet biological weapons program: A history*. Harvard: Harvard University Press.

MacKenzie, D., and G. Spinardi. 1995. Tacit knowledge, weapons design, and the uninvention of nuclear weapons. *American Journal of Sociology* 101(1): 44–99.

Mayor, A. 2003. *Greek fire, poison arrows and scorpion bombs: Biological and chemical warfare in the ancient world*. London: Duckworth.

McLeish, C. 2007. Reflecting on the dual-use problem. In *A web of prevention: Biological weapons, life sciences and the governance of research*, ed. B. Rappert and C. McLeish. London: Earthscan.

McLeish, C., and B. Balmer. 2012. Development of the V-series nerve agents. In *Innovation and security: Preventing the misuse of new biological and chemical technologies*, ed. J.B. Tucker. Cambridge, MA: The MIT Press.

Meselson, M., J. Guillemin, M. Hugh-Jones, A. Langmuir, I. Popova, A. Shelokov, and O. Yampolskaya. 1994. The Sverdlovsk anthrax outbreak of 1979. *Science* 266: 1202–1207.

Miller, J., S. Engelberg, and W. Broad. 2001. *Germs: Biological weapons and America's secret war*. London: Simon and Schuster.

Polanyi, M. 1958. *Personal knowledge: Towards a post-critical philosophy*. Chicago: Chicago University Press.

Price, R. 1997. *The chemical weapons taboo*. Cornell: Cornell University Press.

Rappert, B., and B. Balmer. 2007. Rethinking 'secrecy' and 'disclosure': What science and technology studies can offer attempts to govern WMD threats. In *Technology and security: Governing threats in the new millenium*, ed. B. Rappert, 45–65. Palgrave: Basingstoke.

Schiebinger, L. 2001. *What has feminism done for science?* Cambridge, MA: Harvard University Press.

Selgelid, M. 2007. A tale of two studies: Ethics, bioterrorism and the censorship of science. *Hastings Center Report* 37(3): 35–43.

Sims, N. 1988. *The diplomacy of biological disarmament*. New York: St. Martin's Press.

Sims, N. 2001. *The evolution of biological disarmament*. Oxford: Oxford University Press.

SIPRI. 1971. *The problem of chemical and biological warfare*, 6 vols. Stockholm: Almqvst and Wiksell.

Strauss, H., and J. King. 1986. The fallacy of defensive biological weapons programmes. In *Biological and toxin weapons today*, ed. E. Geissler, 66–73. Oxford: Oxford University Press.

Tucker, J. 1994. Dilemmas of a dual-use technology: Toxins in medicine and warfare. *Politics and the Life Sciences* 13(1): 51–62.

Tucker, J. 2007. *War of nerves: Chemical warfare from World War I to Al-Qaeda*. New York: Anchor Books.

Tucker, J., and E.R. Mahan. 2009. *President Nixon's decision to renounce the US offensive biological weapons program*. Washington, DC: National Defence University Press.

United Nations. 1972. Text of the biological weapons convention. http://www.un.org/disarmament/ WMD/Bio/pdf/Text_of_the_Convention.pdf. Accessed on 22 July 2014.

Vogel, K.M. 2008. Framing biosecurity: An alternative to the biotech revolution model? *Science and Public Policy* 35(1): 45–54.

Walker, J. 2012. *Britain and disarmament: The UK and nuclear, biological and chemical weapons arms control and programmes 1956–1975*. Farnham: Ashgate.

Weber, R. 1997. Manufacturing gender in commercial and military cockpit design. *Science, Technology and Human Values* 23: 235–253.

Wheelis, M., L. Rózsa, and M. Dando (eds.). 2006. *Deadly cultures: Biological weapons since 1945*. Cambridge, MA: Harvard University Press.

Wright, S. 2002. Geopolitical origins. In *Biological warfare and disarmament: New problems/new perspectives*, ed. S. Wright. Lanham: Rowman and Littlefield Publishers.

Chapter 10
Technology and Ecological Values: Confronting *Normal Waste* as Unavoidable Matter in Modern Society

Helena Mateus Jerónimo

If we think about the topic of waste today we have to be aware of its plastic and ambiguous nature, rejecting both the economic view, which sees it as lost and as a negative value, and the socio-cultural view, which associates it merely with fear and repugnance. Waste is not a single unified category which can be understood in linear fashion. Everything comes together in waste in a series of paradoxes: production and consumption; past and future; materials and social sensitivities; economic relations and social and aesthetic relations; value and non-value; repugnance and desire; fear and moral/cultural norms; profit and worthlessness; the useful and the superfluous; excess and loss. It is something omnipresent, fluid and liminal, lying between material object, experience and metaphor. It reveals a great deal about the way we live and about prevailing values, about the economy which produces it, and of our notions of development.

Even though waste is important, and relevant, scholars have paid it little attention, particularly compared to the abundance of research and published work on consumption. The reasons for relegating it to the background are obvious: "waste is society's dirty secret world" (Engler 2004, p. 14). But consumption and waste go together, and are characteristic of societies both past and present. Some writers argue that waste is the key topic, and that rather than talking of "consumer societies" we should perhaps be talking of "rubbish societies," in that rubbish is such an important engine of their growth and transformation as well as for production and consumption (O'Brien 2008, p. 5). Just as the ruins and remains of earlier civilizations give us significant insights into the past, so today's wastes will be the ruins of the future and will reveal a great deal about today's multifaceted meanings and values.

"Normal waste" summarizes in a two-word formula the idea that waste is an inherent condition of a society of widespread production and consumption,

H.M. Jerónimo (✉)
ISEG, University of Lisboa and SOCIUS, Miguel Lupi street, 20, 1249-078 Lisboa, Portugal
e-mail: jeronimo@iseg.ulisboa.pt

© Springer International Publishing Switzerland 2015   183
W.J. Gonzalez (ed.), *New Perspectives on Technology, Values,*
and Ethics, Boston Studies in the Philosophy and History of Science 315,
DOI 10.1007/978-3-319-21870-0_10

something central and unavoidable that spawned social, cultural, economic and technical responses, which in turn shape our history.[1] But because this is a problem which is widespread, permanent and global, with environmental and public health effects on current and future generations, intervention in this field has to be guided by a combination of responsible use of resources, sustainable patterns of production, consumption and development, and the involvement of citizen-consumers.

From these foundations I seek to show, first, that waste as a "problem" is the product of an order of production and consumption increasingly shaped by industrialization and urbanization. This notion was accompanied (or perhaps provoked) by a long drawn-out cultural process in the realm of sensitivities, mentalities and philosophical and medical convictions, which encouraged the sanitizing of public spaces, greater individuality, the refinement of manners and of the sense of smell, and normalization of behaviour. Secondly, I see waste as an environmental issue and as a factor which encourages the search for ingenious technological developments, stimulates international political measures, and involves multiple networks and institutions. The belief that the "problem" of waste can be only fixed technically is pure fantasy. Facing up to the problem effectively will also depend on co-ordinated political approaches and patterns of development which achieve harmonious combinations of the social, the economic and the political. Finally, I analyse waste in the context of its commodification. Since there are many ways of appropriating it, some writers refer to the "redemptive" capacity of waste (Hawkins and Muecke 2003), and others to its "alchemical conversion into value" (O'Brien 2008, p. 5). That which at one stage is a waste product may become a resource in the next.

10.1 Sanitization and Sensitivities

Among the many new and complex issues which emerged with the modern world, clearly related to industrialization and the establishment of large urban metropolises, was the exponential growth of waste and a concern with hygiene in a broad sense.[2] Despite the fact that it has persisted throughout the history of mankind, it was only in this context that waste begun to be considered as a "problem" and no longer as a mere annoyance that could easily be avoided by removing it from the range of human senses. According to the historian Martin V. Melosi (1981), there are two factors in the emergence of this notion of waste as a "problem." On the one hand, it is perceived to be one of the negative and unavoidable by-products of the historic connection between industrialization and urbanization,[3] a kind of "urban

---

[1] This idea, which partially provided the title for this article, overlaps with those of Perrow (1984). O'Brien (2008) also points out that waste is a *normal* and unavoidable constituent of life.

[2] This point of sanitization and sensitivities is discussed in more detail in Jerónimo (2010).

[3] The connection Melosi establishes should be seen in a non-linear way. For example, industrialization in Britain began outside the cities, and the growth of the great European cities in the nineteenth century was only partly due to industrialization. For these reasons, urban disease can only partly

blight," aggravated by limited space, and dense concentration of populations and industries. On the other, it gained public recognition as an environmental issue and a serious danger for human health, as huge quantities of industrial and urban rubbish were far beyond the capacities of traditional collection and disposal practices and so demonstrated the impracticability of the philosophy of "out of sight, out of mind."[4]

Dumping into lakes, rivers, harbours, and even the open sea was still the most common procedure. The environment was regarded "as an abstraction," in so far as air, sunlight, the purity of rivers, "because of their deplorable lack of value in exchange, had no reality at all" (Mumford 1963 [1934], p. 168). Increasing awareness that an "out of sight, out of mind" mentality was no longer appropriate to the circumstances, and of the relationship between waste and diseases, forced local councils to introduce reforms and look for new technical solutions for waste disposal. Sanitation reforms, spearheaded by medical scientists, public health advocates and city officials, were the first steps in the politicisation of waste. This sanitization movement was based on the miasmic theory whereby diseases and epidemics were caused by "bad air," air pollution and poisonous emanations from drains, cemeteries, cesspits, sewers and other foul places. Emphasizing cleanliness and the circulation of air, sanitation reformers demanded that cities undertake the eradication of public nuisances. Edwin Chadwick, Charles Turner Thackrah, Baron Haussmann, Florence Nightingale, George E. Waring, Jr., Piotr Kropotkin, and Ebenezer Howard were some of those who either carried out or inspired sanitation and city reformers.

In a prevailing climate of fear of the resurgence of epidemics, and under the influence of the medical view of the human body, the purpose of the reforms was to plan and organize cities so that they resembled a healthy body, with a clean skin, healthy breathing and fluid circulation. The city was like "a patient on the operating table, to whom the engineers applied a scalpel with no restrictions" (Saraiva 2005, p. 14). So plans were drawn up to take waste away from human contact and human sight; to push waste dumps to the outskirts of towns; to clean the gardens, where previously waste had been collected and stored; to pave all the city's roads (its "arteries"), so that they could more easily be cleaned; to build "veins" of sewers underground; to create large green "lungs," around which the "arteries" would go, so that people could breathe pure air (Sennett 1994, pp. 263–264, and 325; Engler 2004, p. 55).

With the dissemination of the discoveries of Louis Pasteur, Joseph Lister, Robert Koch, among many others, diseases began to be explained by invisible and odourless pathogenic organisms, such as viruses and bacteria. The germ theorists proved their case against the miasmists. The actions taken on the basis of the miasma theory were, however, crucial in terms of cleaning up towns and improving living standards. Hospital hygiene improved greatly, even before the germ theory of disease

---

be blamed on industry, even if it was indirectly responsible in that it made large-scale mass consumption possible and actively encouraged it.

[4] In the English-language literature, we often find the expression *out of sight, out of mind* to express the notion of removing waste far from where it can be seen, smelled or touched by humans.

came to its aid, and even though it did so after sanctioning hygienic practices and justifying them as proof of the validity of the theory.

An early notion of "public service" in the areas of sanitation, public health and housing policy emerged before the end of the nineteenth century.[5] City services were established to provide solutions to collective needs, from water supplies to waste collection, in a sort of "municipal socialism" (Melosi 1981, p. 12; Mumford 1961, p. 476). But what really revolutionised sanitation was the interactive synergy of various technological systems, in particular the creation of an efficient water supply network and the technical solution to the problem of how to remove excrement through sewers (Silva and Matos 2004). Technical efficiency and complementarity, on the one hand, and medical legitimacy, on the other, drove the improvements in hygiene, cleanliness and safety.

Urban waste is no longer directly associated with illness and death. It has begun to be seen as a technical problem requiring effective treatment systems and its own appropriate disposal facilities, managed by specialist technicians. Waste treatment techniques and greater efficiency in waste collection have helped to make waste "invisible," an outcome which is in line with the interests and expectations of capital, industry and business. The symptoms would be dealt with rather than the underlying causes, because it was a rare thing to challenge the basic principles of the market system which consumed resources and produced rubbish (Rogers 2005).

Reflections based on a historic contextualization of waste as a "problem" and as the driving force behind reforms in the field of sanitation, hygiene and urban planning, cannot be dissociated from the empirical evidence of the emergence of new sensitivities, new mentalities and new forms of sociability. The long cultural process of refinement of manners and the rising threshold of sensitivity explains people's relationship to waste, governed by the experience of repugnance and the gradual sanitization and deodorization of homes and public spaces.[6] The concrete examples of that process are a part of what Norbert Elias calls "civilizing" changes (2000 [1939]). Gestures accompany the refinement of modesty, repugnance and intimacy and express the human tendency for increased control over everything which shows up the "animal side," rendering it less visible or confining it to the intimate sphere. People begin to regard so-called "grotesque" behaviour with aversion, and it gradually changes towards a sense of what was called "civilized," leading not only to the emergence of new standards of personal and domestic hygiene

---

[5] In the twenty-first century, sanitary and living conditions for many millions of people in cities of Africa, Asia and Latin America are hardly decent, or even non-existent. Here the problem of what to do with human excrement continues to be a daily concern and a real danger in ecological and public health terms. In this connection see the troubling data contained in Black and Fawcett (2008) and George (2008).

[6] For a deeper understanding of the historical development of infrastructure and mental attitudes, see the monumental work by Philippe Ariès and Georges Duby, *Histoire de la Vie Privée* (1987), which belongs to the French historiographic tradition of the *Annales*. Among numerous examples, we may cite changing attitudes to the kitchen and the bathroom, which until then had been objects of utter indifference and even contempt. On these two types of space, see Wright (1960), Lupton and Miller (1996), and Horan (1996).

and deodorization (Vigarello 1987 [1985]), but also to the spread of these practices to the whole city, in an attempt to maintain equilibrium between the atmosphere and bodily fluids. The French historian Alain Corbin speaks accordingly of the "olfactory revolution" and "the great dream of disinfection." There is nothing too surprising here if we bear in mind that cities in mid-eighteenth century had "the olfactory intensity of an environment of excrement" (1986 [1982]), p. 30).

These changes at the cultural level were to come up against the huge growth of waste which took place after the end of the Second World War, as a result of the advent of mass production and consumption combined with the *throwaway culture*. There was, in the words of Vance Packard (1960), a "planned obsolescence" strategy, a way to shorten life expectancy of products and fuel ever-expanding human needs. Products rapidly became obsolete, as a result of trends in fashion, technological innovation, or planned obsolescence (whether planned or as an accidental effect of technological progress). The outcome of mass consumption was the corresponding mass production of different types of waste and the concern with how to manage them. Consuming and throwing away represented economic prosperity, progress, convenience and freedom. As Susan Strasser explains, in *Waste and Want*, the triumph of disposable products, for example, is due "to their ability to make people feel rich: with throwaway products, they could obtain levels of cleanliness and convenience once available only to people with many servants" (Strasser 1999, p. 9). In this context, the practice of re-use and recycling came to be looked at with indifference, as something obsolete and unnecessary, and as part of the environmentalist activism of the counterculture, rather than as acts of civic, municipal or corporate responsibility with regard to waste (ibid., p. 283).

At the end of the twentieth century, waste and consumption had become radically different things to what they were in earlier historical epochs. Not only were there far more sources of waste–domestic, commercial, industrial, medical, agricultural, construction, and so on–but there had been changes to its physical and chemical nature, comprising metals, plastics, glass, paper, and vegetable matter, often in complex and hard-to-separate combinations as in batteries, cartons and cars. There is also a tendency for the lifetime of "durable" products to be shortened to that of "consumables," and for non-renewable natural resource stocks to be consumed in the same way as renewable production flows.

10.2   Environment and Technology

Despite the fact that it has been a constant in human history, and has been seen as a problem since the advent of industrialization, it was only in the 1960s that waste itself became to be regarded as a global environmental problem. This new approach to waste can be seen in the context of emerging ecological movements and environmental ethics, especially Rachel Carson's pioneering book on the harmful effects on human and animal health of the massive application of synthetic pesticides in agriculture, published in 1962; the alert launched by Club of Rome in 1972[7]; and the UN Conference in that same year ("the Stockholm Conference"), which gave birth to the "Declaration on the Human Environment" and to the UNEP. This took place alongside feelings of ambiguity in relation to the unexpected consequences of technological advances, reinforced by man-made environmental accidents in the fields of technology and energy. Times Beach and Love Canal in USA, and the Seveso dioxin-contaminated waste drums in Europe are waste-related examples of this.[8]

Waste began to be a problem for the environment and for the way we live. Today there seems to be a growing consensus that waste is above all the product of a specific model of development: underlying it is a particular conception of time, as far as the human ability to solve problems is concerned. The threat that waste poses to health and to the environment is not just something which affects the current generation, nor can it be solved by the development model which produced it. On the contrary, waste is a problem because natural resources (oceans and rivers, the soil, air, etc.) are scarce and fragile and because it has long-term effects. In addition, we also have to consider its impact on the quality of urban life, aesthetics, affluence, and technology. Waste policy, as part of a wider strategy either to decrease pollution, to protect the environment, or to bring about technological and industrial change and innovation, began to bring about substantive changes at both the macro level of legislation and the micro level of domestic practices and consumption habits. Public recognition of the potential threats associated with waste lead to resistance to such facilities being located in people's backyards. A disposal approach gave way to a management approach. Waste began to be managed in the light of environmental factors and an ethos of prudent living. Once waste had become a management "problem," international bodies embarked on a great variety of efforts to harmonise legislation and data collection policies. Even so, there are major disparities in waste statistics, arising from countries' different methods and criteria

---

[7] The study, directed by Dennis and Donella Meadows, under the title *The Limits to Growth,* concluded that the limits to planetary growth would be reached within a short period of time (a maximum of 100 years), if levels of industrialization, pollution, food production and exhaustion of natural resources were maintained (Meadows et al. 1972). There were many reactions to what was regarded as mere apocalyptical speculation–all the more so because the study called for neo-Malthusianism, or a theory of zero growth in population and industry, as a solution to the imminent "catastrophe" –, but the report did have the merit of placing the environmental issue on the political agenda and of warning of the urgent need for a slowdown. After this first report, the same team has conducted a 20-year update of the original study and published the results in *Beyond the Limits,* in 1992, in addition to the report entitled *Limits to Growth: the 30-Year Update.*

[8] It is worth recalling the accident at Love Canal (in Niagara Falls, New York) because it was the first hazardous waste disposal case to draw international attention. The canal had originally been built in 1890 by William T. Love (hence its name) to serve as a water channel, but it was never concluded, and ended up being used as a dump for chemical waste from the 1930s to the 1950s. When it became full of waste, it was covered over and sold. Residential houses and a school were built on top of it. It was not long before the inhabitants started to complain of strange smells and health problems. In 1980, the Love Canal area was declared a federal emergency area by President Carter. The fact that this was the first time such a declaration had been made marks the beginning of a time of awareness of the risks associated with hazardous waste disposal (Levine 1982).

used to record information on the production, transportation, treatment and final destination of wastes. This is one of the main barriers to obtaining reliable and comparable data needed to define proper regulatory plans and monitoring of waste products. In addition to this considerable ignorance regarding the nature and precise extent of the problem, there is also ignorance of the amounts of waste which are not declared, are exported or just dumped somewhere illegally.

Despite all the legislative effort, despite the fact that waste recovery (e.g., recycling and re-use) has grown considerably, and despite the fact that many industries now use cleaner technologies, these developments have not been sufficient to reverse the trend for waste production to increase, at a substantially faster rate than economic growth. For example, for 2020, the OECD estimates that waste production will have grown by 45 % compared to 1995.[9]

In the standard pattern of waste management, the priority given to prevention and minimization shows up something which is often overlooked when talking about waste. Besides the financial and environmental costs of waste to society – it has to be collected, treated and disposed of, it also represents a significant loss of raw material and energy resources. In general terms, the productivity/waste coefficient is a good indicator of a society's economic efficiency. The higher the production of waste, the lower the efficiency of the production process, the lower the durability of products and the greater the waste of resources (EEA 1999). That is the reason why the idea of waste as an "inconvenience," easily sanitized provided it was removed from human senses and city centres (that "out of sight, out of mind" attitude which prevailed in earlier times), has given way to a proactive attitude of prevention.

We should not forget, however, that waste management activities are also significant business opportunities. An entirely new industry has developed over the years to deal with the waste problem, and a significant number of waste-related technologies and services, from collection to recycling, are very profitable and can provide new markets. For example, the principles of current international waste management strategy (reduction, recycling and re-use, and improving final disposal and monitoring) require significant investment in technological and scientific innovation. Thus, the ability to recycle is built into some products at the design stage; a less wasteful product design and manufacturing process makes cost savings attainable; and some technological innovations are created specifically to improve the treatment or recovery of waste. In these aims, there is remarkable ambivalence regarding the technological implications. On the one hand, technology itself is responsible for much of the waste production and global pollution. Each technical development, despite its many benefits, has brought an increase in the amounts and types of waste. After the non-degradable waste produced by the iron and steel industries of the early industrial era, plastic, chemical, and pharmaceutical products have given rise to even more waste products which are more toxic and difficult to treat, control, and dispose of. On the other hand, technology is also absolutely necessary for waste prevention and the disposal of pollutants.

---

[9]According to data on the European Union website, http://europa.eu.int/comm/environment/waste/index.htm, accessed on May 2013.

Decisions on waste which affect populations should not be limited to the technical aspects of the problem, nor confined to an elite of experts. Nor should they be subject to processes which rule out all external control and acceptance. Intervention in this area has social, economic, environmental, ethical and political ramifications, including choices concerning resources and models of development, cultural attitudes to consumption and the rights of future generations. However significant the advances in the technology of waste management and disposal, these are mere "remedies." They should not blind us to the seriousness of the "disease."

10.3   Commodity, Art and Habitat

The many shifts in the interpretation of waste have meant that it has always been at the centre of social and economic life. In a more radical perspective, waste can be seen as having the ability to unmask a system which exploits nature, life and human labour and converts them into commodities (Rogers 2005, p. 230). The emergence of corporate environmentalism or "green" (not necessarily environmentally-friendly) industries and products, which do not really question the prevailing capitalist development model, are often viewed sceptically. Production on a large scale, "binge" consumption and waste have been and continue to be vital conditions for capitalism. Without *destruction*, there would be no need for constant *creation*, to borrow Joseph A. Schumpeter's concept of "creative destruction" (Schumpeter 1987 [1942]). As industry expanded into multiplying the number and variety of goods, and particularly into creating incentives for people to desire more of those goods, the avoidance of waste came to be something which would threaten the integrity and foundations of the very system which managed it, thereby provoking its own destruction.

Throwing things away was the engine and driver of further consumption – consumption constantly incentivised, and still today a significant means of social differentiation based on financial status, whereby it is appearance, personal show, the management of impressions and the cult of novelty that count. There are echoes here of the pioneering arguments of Thorstein Veblen (1973 [1899]) on "conspicuous consumption" and "conspicuous waste."[10] This theory was revived, almost a

---

[10] Unlike other economists of his day, who tended to explain wealth creation on the basis of factors of production, Veblen finds the engine of economic development in consumption. He highlights the functions of objects of consumption as measures of value and social differentiation as being more important than their practical and functional status. The concept of conspicuous consumption represented the aristocratic style of life, of the so-called "leisure class," founded on ostentation and wasteful expenditure. Excessive consumption and the waste of time, whether directly or by proxy, was seen as a sign of wealth, social status and prestige, in sum, a symbol of belonging to a privileged group. Veblen does show that consumption by an individual or by a social group is not independent of consumption by others, contrary to what was stated by rational choice theory.

century later, by Jean Baudrillard (1981), according to whom the choice and consumption of objects continues to play a distinctive social role in modern society, and by Pierre Bourdieu (1982 [1979]) in his studies on the different lifestyles of the classes and strategies for social differentiation.

Waste, however, does not let itself be rigidly separated into what is worth something and what is not. The conversion of "waste" into "non-waste" produces significant ambiguity in the categorization of materials as "rubbish," "goods" or "raw materials," and in the way they are regulated. For example, the specific characteristic of the waste treatment, recycling and disposal industries is to work with raw materials which are the waste products of other industries. Since this is a negative-value resource, those industries have a source of revenue where conventional industry has costs. Not only do they get the raw materials they need at zero cost – they are also paid to handle, recycle, or dispose of them. The situation is different when we are dealing with waste products which may appreciate in value (e.g., silver residues in photographic wastes), since the treatment industries involved pay for these wastes, instead of being paid to take them away (Wynne 1987).

Just as waste is recycled for viable industrial purposes, as raw material or source of energy, so art has adopted waste as a material which is marketed as something of aesthetic and monetary value. Although is has existed for centuries, this type of art has always been on the fringe of the fine arts world. It was only with the increasing popularity of environmental movements that it also gained recognition in those markets. Artists' creative freedom and the constant search for new materials have enabled them to realize aesthetic potential in waste products, transforming them by invention, recycling and re-using them. In responding to the challenge of what shocks the eye, the nose and the sense of touch, several artists, sculptors, stylists and photographers have increasingly aestheticized waste, endowing it with ambiguities which lie between ugliness and beauty, transforming aversion into objects of desire. According to art historian Jo Anna Isaak (2001), the paradoxical nature of waste is revealed in an "anti-aesthetic duality," thus illustrating how works of this kind blur the boundaries between that which is aesthetic and that which is not. Thus waste, when it becomes a valued resource, liberates itself from an exclusive connotation with the notions of dirt, chaos, disorder and contagion which anthropological studies showed to be traditionally associated with the breakdown of the established order (Douglas 1966).

Artists have created a new visual culture, which may be interpreted either as a critique of the ostentatious nature of consumption and the utilitarian values of the capitalist economy, or as a particular expression of recycling and environmentalism, or as reflecting anguish at the technological colonisation of the world. The aestheticization of waste may be thought of as a tendency having similar characteristics to many of the avant-garde art movements of the twentieth century (e.g. Arte Povera; the collage work of Picasso or Braque; the re-appropriation of mass-produced and short-lived objects, such as Duchamp's *Bicycle Wheel*). We need only think of contemporary examples such as the photographs of Chris Jordan (e.g., the collection entitled "Intolerable beauty: portraits of American mass consumption") or those of Vik Muniz (as seen in the documentary "Waste Land"); the sculptures of H. A.

Schult (the "army" of Trash People, life-size figures built from rubbish which have been exhibited in various capitals since 1996); the unique luxury (Freitag) handbags, by the Swiss stylists Markus and Daniel Freitag, made entirely from recycled products (used car seat belts, air bags and bicycle inner tubes). The aesthetic appropriation of waste allows us to explore the potential for ambiguity in Elias' theories. It suggests that there is a "dual civilizational mechanism" at work, whereby the repugnant is banished backstage, but at the same time is brought back to the surface by being aestheticized.[11]

For many sections of the population waste is also a habitat. They live off waste, and have themselves become a kind of "social trash." The world which produces various types of waste also produces "human scum," "sub-humans," "sub-citizens," who live in a state of sub-citizenship and marginalisation, in the cracks and margins of the system (Souza 2006) or in an "underworld" as in the novels of Don DeLillo. They live from and with(in) waste. They embody the figure of Georg Simmel's *stranger*, he who is near and far away at the same time. It is relevant in this connection that in earlier periods of history the collection and treatment of rubbish was always assigned to "the socially excluded" (prisoners of war, condemned men, slaves, the hangman's assistant, prostitutes, beggars, etc.)" (Eigenheer 2003, pp. 32–33). In several places, like Accra, the capital of Ghana, and Guiyu, in southern China, or in Jardim Gramacho, a big open-air landfill outside Rio de Janeiro, Brazil, just to mention a few examples, recycling is many families' daily bread. Adults and children smash up obsolete electronic equipment with stones and then burn it, after extracting metals such as silver, gold, palladium, aluminium and copper from processors, chips and connecting pins, also known as "gold fingers." This process tends to release toxic fumes, which cause severe respiratory problems when inhaled.

This dilemma, which has existed for several decades, of the existence of conspicuous consumption, sophisticated technology and environmental pollution, has encouraged the prospects for nonconsumption and alternative technologies. Since the mass consumption society is based on limitless growth, but natural resources and material progress are finite, many writers have highlighted the need to promote forms of growth which are more serene, convivial and sustainable, or even a total break with the notion of growth. Writers like Ivan Illich and Cornelius Castoriadis, and more recently Serge Latouche (e.g. 2007), and others like Mol et al. (e.g. 2009), despite their differences, have called for a reorganisation of modern societies based on ecological rationality, as articulated in concepts such as "degrowth" and "ecological modernization."

---

[11] I would like to express my thanks to Rafael Marques for calling my attention to this point.

10.4   Concluding Remarks

The way we get rid of waste (and appropriate it) amounts to an ethos which is aware of historical contexts and their political, social and intellectual dynamics. Waste products and resource use are extensively tied in with the ends and means of modern societies. Dealing with these issues involves questioning the foundation on which those societies were established in the past and on which they rely in the present.

As we have seen, waste is not "abnormal." Rather, it is ambivalent, and an integral part of social life. The boundaries of waste are hazy, and it has been appropriated in many different ways. Waste is a key mechanism in the maintenance of capitalism. It nourishes an economy which adds commercial value to it, inasmuch as it can be transformed into a source of profit and a market for labour (e.g., in recycling facilities); generate income and resources for certain sections of the population (e.g., the homeless, rag pickers); be reprocessed into fuel (an energy commodity); be converted into an art object (e.g., *arte povera*); be recovered as a historical relic; and have its language appropriated by the rating agencies to classify some countries (junk bond) and to describe the electronic "spam" which invades our e-mail accounts or the "garbage" of newscasts and newspapers. Waste may also lead to technical innovation, or play an important role in changing social values, as has been demonstrated by the change in general habits as a result of recycling.

The fact that we recognize waste as a normal and inevitable condition of our societies does not mean we should condone its limitless production. Many of the technological innovations in waste management and disposal are important, but they arise downstream, and are part of the prevailing technical and scientific matrix, which is often biased in favour of private and corporate interests. This means that those technologies not only adjust to the waste-producing model: they also ratify, justify and legitimate it. The fact that international reports have forecast significant increases in various types of waste, thus signalling the failure of the key objective of reducing waste production, represents in the final analysis the success of a whole network of organisations and practices dedicated to finding value in waste and accelerating the cycle of production, consumption and waste.

Challenging the causes of increased waste production means questioning trends and inclinations which are ingrained in the instrumental culture of modernity itself. Modern society strives for a balance between economic development and environmental protection, finding a threshold that reconciles the inevitable production of waste with a commitment to ecological sustainability. The depletion of natural resources that may not be renewable, and the (often related) by-production of hazardous waste, is an increasingly important focus of long-running debates regarding conflict between state regulation and market forces, between individual action and collective consequences, and between the practical and the ethical impact of new or newly mass-consumed technologies.

References

Ariès, Ph., and G. Duby. 1987. *Histoire de la Vie Privée*, 5 vols. Paris: Seuil.

Black, M., and B. Fawcett. 2008. *The last taboo: Opening the door on the global sanitation crisis*. London: Earthscan.

Bourdieu, P. 1982 [1979]. *La Distinction: Critique Social du Jugement*. Paris: Minuit.

Baudrillard, J. 1981. *For a critique of the political economy of the sign*. St. Louis: Telos Press.

Carson, R. 1962. *Silent spring*. Boston: Houghton Mifflin.

Corbin, A. 1986. *Le Miasme et la Jonquille: l'Odorat et l'Imaginaire Social XVIIIe- XIXe Siècles*. Paris: Flammarion.

Douglas, M. 1966. *Purity and danger*. London: Routledge and K. Paul.

EEA (1999) *Environment in the European Union at the turn of the century*. European Environmental Agency, Copenhagen

Eigenheer, E.M. 2003. *Lixo, Vanitas e Morte*. Niterói: EdUFF.

Elias, N. 2000 [1939]. *The civilizing process. Sociogenetic and psychogenetic investigations*. Oxford: Blackwell.

Engler, M. 2004. *Designing America's waste landscapes*. Baltimore: The Johns Hopkins University Press.

George, R. 2008. *The big necessity: The unmentionable world of human waste and why it matters*. London: Metropolitan Books.

Hawkins, G., and S. Muecke. 2003. *Culture and waste: The creation and destruction of value*. Lanham: Rowman and Littlefield.

Horan, J.L. 1996. *The porcelain god: A social history of the toilet*. Secaurus: Crol.

Isaak, J.A. 2001. Trash: Public art by the garbage girls. In *Gendering landscape art*, ed. S. Adams and A. Gruetsner Robins, 173–185. New Brunswick: Rutgers University Press.

Jerónimo, H.M. 2010. *Queimar a Incerteza: Poder e Ambiente no Conflito da Co-Incineração de Resíduos Industriais Perigosos*. Lisbon: Imprensa de Ciências Sociais.

Latouche, S. 2007. *Petit Traité de la Décroissance Sereine*. Paris: Mille et Une Nuits.

Levine, A.G. 1982. *Love canal: Science, politics, and people*. Lexington: Lexington Books.

Lupton, E., and J.A. Miller. 1996. *The bathroom, the kitchen and the aesthetics of waste*. Dalton: Princeton Architectural Press.

Meadows, D.H., D.L. Meadows, J. Randers, and W.W. Behrens III. 1972. *The limits to growth*. New York: Universe Books.

Melosi, M.V. 1981. *Garbage in the cities: Refuse, reform, and the environment, 1880–1980*. College Station: Texas A and M University Press.

Mol, A.P.J., D.A. Sonnenfeld, and G. Spaargaren. 2009. *The ecological modernization reader: Environmental reform in theory and practice*. London/New York: Routledge.

Mumford, L. 1961. *The city in history: Its origins, its transformations, and its prospects*. London: Secker and Warburg.

Mumford, L. 1963 [1934]. *Technics and civilization*. New York/London: Harcourt, Brace and World.

O'Brien, M. 2008. *A crisis of waste? Understanding the rubbish society*. London: Routledge.

Packard, V. 1960. *The waste makers*. London: Penguin.

Perrow, Ch. 1984. *Normal accidents: Living with high-risk technologies*. New York: Basic Books.

Rogers, H. 2005. *Gone tomorrow: The hidden life of garbage*. New York/London: The New Press.

Saraiva, T. 2005. *Ciencia y Ciudad: Madrid y Lisboa, 1851–1900*. Madrid: Ayuntamiento de Madrid.

Schumpeter, J.A. 1987 [1942]. *Capitalism, socialism, and democracy*. London: Unwin.

Sennett, R. 1994. *Flesh and stone: The body and the city in western civilization*. London/Boston: Faber and Faber.

Silva, A. Ferreira da, and A. Cardoso de Matos. 2004. The networked city: Managing power and water utilities in Portugal, 1850s–1920s. *Business and Economic History*. On-line, http://www.thebhc.org/publications/BEHonline/2004/daSilvaMatos.pdf. Accessed on June 2013.

Souza, J. 2006. *A Construção Social da Subcidadania: para uma sociologia política da moderni-dade periférica*. IUPERJ, Rio de Janeiro, Belo Horizonte: Editora UFMG.

Strasser, S. 1999. *Waste and want: A social history of trash*. New York: Metropolitan Books.

Veblen, T. 1973 [1899]. *The theory of the leisure class*. Boston: Houghton Mifflin.

Vigarello, G. 1987 [1985]. *Le Propre et le Sale: L'Hygiène du Corps depuis le Moyen Age*. Paris: Seuil.

Wright, L. 1960. *Clean and decent: The fascinating story of the bathroom and water closet*. New York: Viking.

Wynne, B. 1987. Hazardous waste: What kind of issue? In *Risk management and hazardous waste: Implementation and the dialectics of credibility*, ed. B. Wynne, 45–83. Berlin/New York: Springer.

Epilogue
Towards a New Scenario in the Relations Between Technology, Values, and Ethics[1]

Amanda Guillan

Undoubtedly, technology has experienced an intense development in recent years. This development has repercussions both on philosophy and on society in many ways. In the first case, this can be noticed in the huge amount of recent publications related to philosophy of technology, but also in the incidence in other philosophical branches (such as anthropology, logic, etc.).[2] Meanwhile, in the second case, the impact is even more noticeable than in the intellectual realm. De facto, contemporary society has change through technological artifacts that directly affect people's life. A very good example is in the area of information and communication technologies [ICTs], which has a direct effect on the ordinary citizens and their daily interactions with other individuals.

Within this context, a crucial issue is the analysis of the relations between technology, values, and ethics. This topic, which is the focus of attention of the present volume, is especially relevant for philosophical reasons but also for social purposes. The common ground for the analysis, which is assumed by the papers that compound *Technology, Values, and Ethics*, is the acceptance of technology as value-laden. Thus, technology can no longer be seen as a value-free undertaking, because

---

[1] This paper is supported by the Program FPU of the Spanish Ministry of Education, Culture and Sport.

[2] The philosophical study of the human being has been enriched by the comparison with technological devices such as robots; the logical analysis is extended through the problems that come from the uses of computers and the need for logical basis for computer sciences, etc.

A. Guillan
Faculty of Humanities, University of A Coruña, Dr. Vazquez Cabrera, w/n,
15.403 Ferrol, A Coruña, Spain
e-mail: a.guillan@udc.es

© Springer International Publishing Switzerland 2015
W.J. Gonzalez (ed.), *New Perspectives on Technology, Values, and Ethics*, Boston Studies in the Philosophy and History of Science 315, DOI 10.1007/978-3-319-21870-0

it is a human endeavor that, among the possibilities available, freely chooses its objectives, undertakings and products.[3]

From this perspective of human undertaking—complementary to the vision of technology as knowledge and as artifact—there is an analysis of the role of values, in general, and of ethical values, in particular. This involves taking into account the internal values as well as the external ones. Both are needed here: the internal values of technology are related to the technological undertaking itself (i.e., the aims, processes, and results of this undertaking), and the external values are linked with the context of technology (i.e., social, cultural, economic, ecological, etc.).

As Wenceslao J. Gonzalez points out in the first chapter of the volume, axiology of technology has a twofold character. On the one hand, there is the *descriptive* side, which is focused on the values that actually intervene in technology; and, on the other, there is a *prescriptive* facet, which involves the reflection on the values that should have a leading role in technological activity (see Gonzalez 2016). Certainly, the chapters of this book have paid attention to both sides—descriptive and prescriptive—of the philosophical reflection about technology and its relations to values, in general, and ethical values, in particular.

A Holistic Approach to Technology, Values, and Ethics

Following these coordinates of the philosophical analysis, the first part of the book is devoted to "New perspectives on technology, values, and ethics." In my judgment, this part of the volume makes quite an interesting point, which actually contributes to a new scenario regarding the axiology of technology as well as to the ethics of technology. There are several reasons to defend this view:

(i) The relevance of values in technology is emphasized, both for its structure and for its dynamics, as well as the necessity of taking into account both the internal and the external values of technology. This kind of analysis leads to a wide framework—of holistic vision—for the role of values in technology (and, among them, ethical values) [see Gonzalez 2016]. Certainly, such an account allows us to overcome fractional orientations in this regard, such as those views merely focused on "internal" values or just on "external" values alone.[4]

(ii) Concerning the specific realm of engineering, the relations between internal and external values have been analyzed in the book. The study has pointed out that, although it is possible to distinguish between the internal dimension and the external facet of values, they cannot be strictly separated.[5] Even more, external values can help us to provide an account for the internal values, and they can also be internalized in engineering practice (cf. van de Poel 2016).

---

[3] In this regard, there is an interaction between scientific creativity and technological innovation, which is analyzed in Gonzalez (2013b).

[4] This problem of the factional orientations has been pointed out with respect to axiology of science in Gonzalez (2013a, b).

[5] About the holism of values, see Rescher (1993), and (1999).

(iii) Along similar lines of a holistic vision of the values, the analysis of the information and communication technologies (ICTs)—in particular, of the Internet—has shown how the aims, processes, and results of the ICTs are related with the environment (social, cultural, economic, etc.). This can be detected not only when the external values are considered, but also when the study is focused on internal values such as accessibility, versatility, efficacy and efficiency (cf. Neira 2016).

Technological Rationality and Responsibility

De facto, these new perspectives on technology, values, and ethics assume the existence of a *technological rationality*, which is different from a scientific rationality.[6] This leads to another main line of research (the second important point that is made in the present volume): the study of the problem of rationality and responsibility in technology. On the one hand, technology is a rational activity which involves aims, processes, and results; and, on the other, it is developed by human agents who have to make decisions, so the issue of responsibility should also be taken into account.

1. It seems clear that instrumental rationality is a key factor in technology, insofar as it has to do with the selection of adequate means to achieve certain aims. Indeed, the specific technological knowledge has been frequently characterized as a *know-how*, and pays special attention on how to make the artifact. Thus, there is an instrumental component that is a key factor when technology is analyzed. From this point of view, values such as efficacy or efficiency are especially relevant to characterizing what technological rationality consists of (cf. Mitcham 2016).

2. But technological rationality also involves an evaluative sphere, so the aims should be chosen according to what is *preferable*, and not merely taking into account what is actually preferred (see Gómez 2016). Thus, technological rationality has not merely an instrumental dimension: it also has a rationality connected with the realm of values, because technological ends and means should be evaluated according to criteria. From this perspective, it seems clear that ethical values have also a relevant role with respect to the selection of the aims and means of the technological activity.

3. Besides the evaluation of the aims and processes, technological results—the products—should be evaluated as well. This evaluation is made according to values, both internal and external. Among those values, there are ethical ones, which are endogenous and exogenous: "There are ethical values endogenous to technology, insofar as it is a free human activity, and there are also exogenous values to the aims, processes, and results of technology, because it is a human undertaking developed in a social milieu" (Gonzalez 2016, p. 5).

4. Aims and processes of technology are relevant, but above all results of technological undertaking can directly or indirectly affect people, society, and the

---

[6] Regarding scientific rationality and technological rationality, cf. Gonzalez (1998). On the differences between science and technology, see Gonzalez (2005) and (2013b).

environment. This repercussion of technology can be either positive or negative. In this regard, a problem arises about who should intervene in the decision-making process related to technological undertaking. It is also a key issue for the question of responsibility in technology. The papers of this book have pointed out the necessity of a wide participation, mainly in the context of democratic societies. This involves several aspects:

(a) Technologists, in general, and engineers, in particular, should take into account ethical criteria when they consider the possible technological alternatives. So, besides the technological expertise with regard the aims and processes of the technological undertaking, there is an ethical component, not only exogenous (related to social pressure), but also endogenous (the ethical evaluation, above all, of aims and means). In this regard, deontological codes are relevant.

(b) Insofar as the aims, processes and results of technological endeavor are related with the environment (social, cultural, economic, ecological, etc.), the specific technological knowledge cannot be enough to guarantee a rational decision-making of the agents. Thus, in order to develop deontological codes for the engineers, technologists should be open to cooperating with other professions that have expertise in different value domains (physicians, psychologists, lawyers, sociologists, philosophers, etc.) [cf. Niiniluoto 1997].

(c) Additionally, citizens themselves, as users of many technological devices, are responsible for the use of technological products. This is clear in the case of "dual-use" technologies, i.e., those technological devices that can be used either for benign or malign purposes (cf. Balmer 2016, especially, pp. 172–173). In the case of online technologies, Juan Bautista Bengoetxea (2016) has shown the necessity to make codes to assist us to make decisions and demand responsibilities from users of those technologies.

(d) Moreover, society as a whole also has a role in the decision-making processes related to technology. As has been pointed out in the present volume, society has the right to expect reasonable ethics of technology (cf. Gonzalez 2016), but it has also the right to cooperate in the development of ethics of technology. Even more, the participation of society itself is needed to develop rational technological policies in democratic societies (this can be seen, for example, in the case of biotechnologies) [cf. Bellver 2016].

The Problem of Risk

Certainly, a key issue regarding technological policy is risk management. This is the focus of attention in the third part of the book, which is devoted to the analysis of the problem of risk with respect to technological development (cf. Rodríguez 2016; Balmer 2016; and Jerónimo 2016). Obviously, technological innovation can be seen as a crucial factor for progress. Moreover, it is a touchstone for economic growth, and contributes to improving our quality of life. But it also happens that technology can be a source of potential health and environmental hazards.

Consequently, when the relations between technology, values and ethics are considered, one of the most relevant issues is how to avoid or, at least, reduce technological risks. In this regard, besides the philosophical reflection on risks—that according to the present volume is developed from an axiological perspective—, there is also a social concern about technology and risks, insofar as technological artifacts can directly affect both people and the environment.

(I) When technological risks are analyzed, the dynamic trait of the axiology of technology is crucial. It has to do with the teleological character of technological undertaking. This undertaking involves the attention to aims, processes, and results, where there is also the problem of the consequences that those results can have. This perspective also comprises a factor of variability, insofar as values can be diverse from one place to another and from a historical moment to another (cf. Gonzalez 2016).

(II) From the dynamic trait, it is more obvious the relevance of the external values. In fact, external values (social, cultural, economic, ecological, etc.)—and, among them, exogenous ethical values—have a leading role when the problem of risk is considered. Insofar as they involve a prescriptive dimension, these external values should be taken into account in the reflection on the limits of technology, especially when there are risks for society (cf. Gonzalez 2016).

(III) This is so because technological development takes place in the context of democratic societies, where the well-being of their citizens should be promoted. From this perspective, it seems clear that citizens themselves have the right to contribute to technological policy, especially in those cases where technology can involve potential risks. Even more, this participation seems to be a requirement for the development of rational technological policies for risk management.

To sum up, there is a new scenario in the relations between technology, values, and ethics due to factors intertwined. On the one hand, the strong development of technology, in general, and information and communications technologies, in particular, has a direct repercussion on the individuals and the societies. On the other hand, there is an intense philosophical interest in values in technology, in general, and ethical values, in particular. Thus, the idea of technology as value-laden and novel ideas on the role of ethical values in technological undertakings give central elements to the new scenario.

References

Balmer, B. 2016. The social dimension of technology: The control of chemical and biological weapons. In *New perspectives on technology, values, and ethics: Theoretical and practical discussions*, Boston studies in the philosophy of science, ed. W.J. Gonzalez, 167–182. Dordrecht: Springer.

Bengoetxea, J.B. 2016. Knowledge and moral responsibility for *online* technology. In *New perspectives on technology, values, and ethics: Theoretical and practical discussions*, Boston studies in the philosophy of science, ed. W.J. Gonzalez, 89–103. Dordrecht: Springer.

Bellver Capella, V. 2016. Biotechnology, ethics and society: The case of genetic manipulation. In *New perspectives on technology, values, and ethics: Theoretical and practical discussions*, Boston studies in the philosophy of science, ed. W.J. Gonzalez, 123–143. Dordrecht: Springer.

Gómez, A. 2016. Risk, uncertainty and the dimension of technological rationality. In *New perspectives on technology, values, and ethics: Theoretical and practical discussions*, Boston studies in the philosophy of science, ed. W.J. Gonzalez, 105–121. Dordrecht: Springer.

Gonzalez, W.J. 1998. Racionalidad científica y racionalidad tecnológica: La mediación de la racionalidad económica". *Agora* 17(2): 95–115.

Gonzalez, W.J. 2005. The philosophical approach to science, technology and society. In *Science, technology and society: A philosophical perspective*, ed. W.J. Gonzalez, 3–49. A Coruña: Netbiblo.

Gonzalez, W.J. 2013a. Value ladenness and the value-free ideal in scientific research. In *Handbook of the philosophical foundations of business ethics*, ed. Ch. Lütge, 1503–1521. Dordrecht: Springer.

Gonzalez, W.J. 2013b. The roles of scientific creativity and technological innovation in the context of complexity of science. In *Creativity, innovation, and complexity in science*, ed. W.J. Gonzalez, 11–40. A Coruña: Netbiblo.

Gonzalez, W.J. 2016. On the role of values in the configuration of technology: From axiology to ethics. In *New perspectives on technology, values, and ethics: Theoretical and practical discussions*, Boston studies in the philosophy of science, ed. W.J. Gonzalez, 3–27. Dordrecht: Springer.

Jerónimo, H. 2016. Technology and ecological values: Confronting *normal waste* as unavoidable matter in modern society. In *New perspectives on technology, values, and ethics: Theoretical and practical discussions*, Boston studies in the philosophy of science, ed. W.J. Gonzalez, 183–195. Dordrecht: Springer.

Mitcham, C. 2016. Rationality in technology and in ethics. In *New perspectives on technology, values, and ethics: Theoretical and practical discussions*, Boston studies in the philosophy of science, ed. W.J. Gonzalez, 63–87. Dordrecht: Springer.

Neira, P. 2016. Values regarding results of the information and communication technologies: Internal values. In *New perspectives on technology, values, and ethics: Theoretical and practical discussions*, Boston studies in the philosophy of science, ed. W.J. Gonzalez, 47–60. Dordrecht: Springer.

Niiniluoto, I. 1997. "Límites de la Tecnología". In González, W.J. (ed.), *Progreso científico e innovación tecnológica*. Monographic issue of *Arbor* 157(620):391–408.

Rescher, N. 1993. *A system of pragmatic idealism. Vol. II: The validity of values: Human values in pragmatic perspective*. Princeton: Princeton University Press.

Rescher, N. 1999. *Razón y valores en la Era científico-tecnológica*. Barcelona: Paidós.

Rodríguez, H. 2016. Risk and trust in institutions that regulate technology: Challenges for a socially legitimate risk analysis. In *New perspectives on technology, values, and ethics: Theoretical and practical discussions*, Boston studies in the philosophy of science, ed. W.J. Gonzalez, 147–166. Dordrecht: Springer.

van de Poel, I. 2016. Values in engineering and technology. In *New perspectives on technology, values, and ethics: Theoretical and practical discussions*, Boston studies in the philosophy of science, ed. W.J. Gonzalez, 29–46. Dordrecht: Springer.

Index of Names

© Springer International Publishing Switzerland 2015
W.J. Gonzalez (ed.), *New Perspectives on Technology, Values,
and Ethics*, Boston Studies in the Philosophy and History of Science 315,
DOI 10.1007/978-3-319-21870-0

Subject Index

A

Accessibility, 42, 47–50, 58, 59, 96, 199
Actor-network theory, 175
Adaptive preferences, 39
Advocate engineer, 81
Aesthetic potential in waste products, 191
Aesthetics, 5, 6, 8–10, 13–15, 42, 43, 47, 49, 73, 183, 188, 191, 192
Agency, 57, 70, 73, 74, 83, 93–94, 96, 97, 100, 110, 115, 152, 156, 193
Agent, 5, 9–13, 16, 17, 19, 64, 69, 70, 73–75, 82, 84, 93–100, 106, 126, 168–175, 180, 199, 200
Agri-food biotechnology, 147–149, 156, 157, 161, 162
Aims, 4–9, 12–14, 17–20, 32, 37, 38, 40, 42, 43, 47–49, 51–53, 56–58, 66, 70, 76, 81, 94, 96, 105, 113, 115, 124, 135, 139, 149, 158, 171, 173, 189, 198–201
Aims, processes, and results, 5, 9, 13, 18, 19, 53, 198, 199, 201
Ambiguous nature, 183
Anatomy, 78
Anthrax, 167, 172, 176, 177
Anthropology, 91, 191, 197
Application of science, 12
Applied science, 5, 7, 12, 15
Arbiter engineer, 81
Architecture, 13, 18, 42, 49, 56, 68
Artifacts, 4, 5, 8–10, 13–15, 17–19, 30–32, 34, 48, 56, 66, 72, 78, 119, 175, 197–199, 201

B

Assessment, 15, 16, 56, 72, 74, 76, 81, 94, 95, 97–99, 108, 109, 112–114, 117, 119, 131, 148, 149, 153, 158–161
Assessment of technology aims, processes, and results, vii, 5, 9, 13, 18, 19, 198, 199, 201
Assisted reproductive technologies, ix
  first test-tube baby, 125
  freezing eggs and ovarian tissue, 126
  global business, 126
  health problems, 131
  infertility, 125, 126, 131
  intracytoplasmic sperm injection (ICSI), 125–126
  preimplantation genetic diagnosis, 126
  reproduction tourism, 134
  in vitro fertilization, 125, 126, 130, 131
Axiological analysis, vi, 3
Axiology
  of the agents developing technology, 12
  descriptive aspect of, 12
  prescriptive aspect of, 12, 15
  of science, 12, 198
  of a specific technology, 12
  of technology, 4, 11–16, 198, 201
  of technology in general, 12

B

Bacillus thuringiensis (Bt), 152
Biological warfare (BW), 167–169, 171, 172, 177–180
Biological weapons, 167–180

© Springer International Publishing Switzerland 2015
W.J. Gonzalez (ed.), *New Perspectives on Technology, Values, and Ethics*, Boston Studies in the Philosophy and History of Science 315,
DOI 10.1007/978-3-319-21870-0

.

Printed in the United States
By Bookmasters